Waste Management for Small Towns in Developing Countries

This book provides a comprehensive technical overview of waste management strategies tailored for small towns, emphasising innovative approaches to integrated waste management. It explores methods to transform various types of waste into value-added products, fostering circular economy practices at both micro and macro levels. The text examines the implementation of waste management solutions across different countries, addressing market development, pricing strategies, and financial viability to achieve sustainability goals. Key topics include municipal solid waste management, faecal sludge treatment, sewage treatment, and grey water management, supported by case studies and practical examples.

Features:

- Reviews integrated waste management systems for small towns with a focus on circular economy principles.
- Highlights indigenous and sustainable waste management practices.
- Explores innovative techniques such as degradable waste segregation and manufacturing value-added products from waste components.
- Covers municipal solid waste management, faecal sludge treatment, sewage treatment, and grey water management, particularly in developing countries.
- Discusses project management aspects and financial implications of waste management initiatives.
- Includes numerical examples, practical problems, and case studies from various countries to illustrate real-world applications.

This book is aimed at graduate students and researchers in environmental, chemical, and civil engineering.

Waste Management for Small Towns in Developing Countries

G. Anand, B. Lakshmikutty, and V. Sayi

CRC Press
Taylor & Francis Group
Boca Raton London New York

CRC Press is an imprint of the
Taylor & Francis Group, an informa business

Designed cover image: Shutterstock

First edition published 2026
by CRC Press
2385 NW Executive Center Drive, Suite 320, Boca Raton FL 33431

and by CRC Press
4 Park Square, Milton Park, Abingdon, Oxon, OX14 4RN

CRC Press is an imprint of Taylor & Francis Group, LLC

ISBN: 9781041091745 (hbk)
ISBN: 9781041095637 (pbk)
ISBN: 9781003650737 (ebk)

DOI: 10.1201/9781003650737

Typeset in Times
by codeMantra

Contents

Foreword

I have gone through the case studies discussed in this book. All the case studies explained in this book had resulted in either commercial benefits or economic benefits over and above having a very positive impact on the society. Ponnampatty Town Panchayat has been developed as a model case of achieving zero-cost operation by ensuring a revenue income to meet the Opex. Ketti and Adigaratty were developed as sources of cost-effective fortified manure which benefited the farmers. The Shrirampur Grey Water treatment plant eliminated public nuisance as well as ensured high-quality treated water for irrigation using low-maintenance technologies. Many of the case studies discussed are FINISH projects and the authors had been associated with FINISH Mondial and FINISH Society India in various ways, and the first author, Anand G had been a consultant and advisor since inception. These case studies indicate how waste management problems in small towns can be effectively managed.

I am certain that this book will be an excellent reference book for students, teachers, sociologists, NGOs, and government staff interested or involved in waste management. I wish the book all success.

Abhijit Banerji
Country Coordinator FINISH Mondial
Member Secretary &Director of the Finish Society
Lucknow, India
July 7, 2025

Preface

Waste Management is an issue which affects the entire society, and naturally, many are interested in it. There are government departments, politicians, NGOs and some members of the general public who get involved in Waste Management. Irrespective of the good intentions and high level of commitment of such persons, there often would be a situation in which they are at a loss to find the way forward. Although there are a number of books available on environmental engineering or waste management, there is a lack of books focused on all kinds of wastes to be managed in a small-town environment. This book is written based on experience in the field and will cater to all types of audiences like practitioners, NGOs, administrators, and even interested general public. Student community can use this book as a reference. This book is a compilation of the results from field experiences of a quarter century which can aid those who work in the field of waste management to identify the core issues, define the objectives and find the contextually most appropriate set of solutions. The axiom of the authors that 'the solution to a problem is the raison d'être of the problem' is at the heart of the approach to the issue of waste management.

Waste management is basically a multi-dimensional activity. There are mutually exclusive components in waste management while some activities can be clubbed together. The approach to holistic waste management depends on the size of the areas under consideration. Large towns and cities have many technologies available for cost-effective waste management while small towns and villages find handling their wastes challenging. Methods like composting can be and have been decentralised effectively when it comes to degradable municipal solid wastes. But they constitute only a part of the several types of waste in the locality. Further, the financial implications of waste management can be daunting for small towns and villages since the economies of scale become critical.

A large segment of the population in developing countries are living in small towns, and due to the fast pace of development, the number of small towns is ever increasing. Urbanisation is a fact of life, and the problems accompanying it are also. Municipal waste management, sewage treatment, faecal sludge management and grey water management are pressing issues which need appropriate solutions to keep the towns clean and disease free. Preserving the lakes, keeping them clean, is very important, as they are the source of groundwater, which is the elixir of life. Governments are spending very large amounts for providing safe drinking water, and yet there is always a water crisis. As mentioned earlier, this book is for all those who are interested in waste management as it suggests how to approach the problem, what the solution can be and how to identify the path to the solution without going deep into the technical and mathematical aspects. This book will be useful to students of undergraduate and graduate studies in the branches of Environmental Engineering, Chemical Engineering, Civil Engineering, Biotechnology, and allied branches as a textbook as well as a reference book.

Since most of the contents of this book are drawn from field experience, some case studies have also been included. Wherever privacy needs to be respected, the same has been factored while preparing this book.

Care has been taken to avoid errors, but to err is human. Any corrections and suggestions for improvement are most welcome

G. Anand
B. Lakshmikutty
V. Sayi

About the Authors

G. Anand graduated in Chemical Engineering from Government Engineering College, Trichur, India, and pursued a career in the manufacturing industry for 13 years before switching to his college dream of Environmental Consultancy. He had been a Process Engineer in Kerala Soaps and Oils Limited, Calicut, India, after which he took up a job as Factory Manager in Savonnerie Indubu, Bujumbura, Burundi. He did a post-graduation in Environmental Science and Industrial Pollution Management before switching careers. In 1995, he founded the consulting firm 'The Solution Centre' in Cochin, India. He was the CEO and Proprietor of The Solutions Centre till he handed over the firm to a successor in 2020. During the 25 years of active consultancy work, his field of activities included Water Supply and Treatment, Sanitation and Waste Management. His clients included various International Non-Governmental Organisations, Not for Profit Organisations, various UN organisations including United Nations Industrial Development Organization, regional NGOs and NPOs, Local self-governments and private enterprises. His work has taken him to several South Asian and African countries. He has conducted several training programmes on Sanitation and Waste Management in India, Sri Lanka, Bangladesh, Ghana, Ethiopia, Kenya and Uganda. Post retirement from The Solutions Centre, he works more as an advisor for some NGOs, NPOs, and some private sector firms and government organisations. He had written/coauthored several manuals and guides on Sanitation and Waste Management for the NGO 'FINISH Mondial/FINISH Society/FINISH INK' and a handbook on Hospital Waste Management in Sri Lanka.

B. Lakshmikutty is a Professor of Chemical Engineering, retired as the Head of the Department, from Government Engineering College, Thrissur. She graduated in Chemical Engineering from Government Engineering College, Thrissur. She was a Research Scholar in the Indian Institute of Science, Bangalore, and later did her M. Tech in Chemical Process Control from the University of Calicut and did her doctorate in the Department of Polymer Science and Rubber Technology, Cochin University of Science and Technology. She worked as Process Engineer in Fertilisers and Chemicals Travancore Ltd, a premier fertiliser company, before joining the teaching profession. She has authored/coauthored research papers, articles and two textbooks 'Stoichiometry and Process Calculations' and 'Mass Transfer Operations Theory and Applications'. She was an expert member in government panels for solving environmental issues related to industrial pollution control. Presently doing academic consultancy work and consultancy in waste management. Post retirement, she associates with FINISH Society, Lucknow, in the field of Waste Management. She has 30 years of teaching experience and is an active member of several professional bodies including Indian Institute of Chemical Engineers (IIChE) and Indian Society for Technical Education (ISTE).

V. Sayi is presently working as a Solid Waste Management Expert in KPMG, which is Project Management Consultant to the 'Kerala solid waste management project' of the Government of Kerala. He graduated in Chemical Engineering from Government Engineering College, Trichur, India, and pursued a career in manufacturing industry in various capacities in Kerala State Drugs and Pharmaceuticals Ltd., Alappuzha, Kerala Soaps and Oils Ltd., Calicut, and Premier Morarji Chemicals Ltd., Alappuzha, before joining as Environmental Engineer in Kerala State Pollution Control Board. He has 27 years of experience in the Department in various capacities and was a senior expert in solid waste management before moving as consultant in Kerala Suchitwa Mission and now working with KPMG. He was a solid waste management consultant to FINISH Society (NGO), Lucknow, for a brief period. As Senior Environmental Engineer, he was working for capacity building of technical staff, imparting training, preparing guidelines and manuals for all types of waste management issues. He was providing guidance and instructions to local bodies in the handling and management of municipal solid waste, hazardous waste, bio-medical waste, plastic waste, E-waste, etc. He was Technical Advisor in publishing Government documents on biogas plants and aerobic composting methods. He provided technical support for construction projects for scientific landfills for solid waste management in Local Self-Government Institutions. He conducted a number of studies like water quality status of important lakes in the state for lake rejuvenation. He was responsible for monitoring noise level and ambient air quality status of industrial, commercial and residential areas. He scrutinized Environmental Impact Assessment Reports submitted by various industries, as part of implementation of Pollution Control Acts and Rules and was responsible for appraising Environmental Audit Reports submitted by various industries as part of the statutory obligation. He had attended various training programmes and presented papers in several national and international forums.

Acknowledgements

This book was made possible largely due to the field experience acquired over 25 years in waste management. The authors are thankful to the following persons for their association in the field of waste management. Many of them are having direct connections with the first author: Valentin Post, WASTE urban advisors/FINISH Mondial, the Netherlands, many of the key projects mentioned in the case studies are in association with him; Abhijit Banerji, Country Coordinator FINISH Mondial India and Member Secretary and Director of FINISH Society with whom all the three authors had been associated in varying capacities; Saurabh Agnihotri, Chief Operating Officer, FINISH Society, Lucknow, for the works in Dungarpur, Rajasthan and K.K. Gupta, former Municipal Chairman, Dungarpur Municipality, a visionary leader who opted for a Sanitary Landfill in a small town along with arrangements for collection of household waste in segregated form; N K Perumal, Director Rural Development Organisation, Bicketty, Nilgiris, Tamil Nadu for the works in Ketti and Adigaratty Panchayat and Chamaraj Tea Estate; K.V. Emmanuel, Environmental Expert, who was part of the UNIDO Regional Asia programme for pollution control in Tanneries and currently Executive Director of Indian Leather Industry Foundation; Dr Nimal Premathilaka, the former Public Health Inspector of Balangoda Urban council, Sri Lanka and currently Technical Specialist World Vision, PHINLA; Nimal Gamin Weerasinghe, former Mayor Balangoda Urban Council, a politician with a remarkable vison for the future of his town and commitment in implementing cost-effective and viable waste management schemes and rejuvenation of water bodies; Ms Pamela Bundi, Country Coordinator FINISH Mondial, Kenya; Dr Sakul Ameedu, who was Executive Officer in both Kattuputhur Town Panchayat and Ponnampatty Town Panchayat and who was committed in implementing schemes with techno economic viability for keeping the towns clean; Padmashri. Dr M. Subburaman, the association with whom gave several opportunities to work with various local self-governments and to get an insight into his very highly committed NGO; K.K. Rajesh, CEO & Managing Partner, RPS Water Technologies and Steby Justin, Chief Technical Officer, RPS Water Technologies, Kochi, for sharing some pictures from their projects for use in this book.

The authors would like to gratefully acknowledge the support of the family members, especially Aysvarya Anand for the inspiration, Ajeya Anand for technical support and Aparna Narayanan for moral and technical support and for being there whenever needed. We gratefully acknowledge the help from the Commissioning editor Dr Gagandeep Singh, PhD, and team of the Publishers, Taylor and Francis.

Abbreviations

ABR	Anaerobic Baffled Reactor
ACF	Activated Carbon Filter
AS	Air Scrub
ASP	Activated Sludge Process
BOD	Biochemical Oxygen Demand
BOO	Build Own Operate
BOOT	Build Own Operate Transfer
BoQ	Bill of Quantities
BS	Broken Stone
BOT	Build Operate Transfer
Capex	Capital Expenditure
CIP	Chemical Clean-in-Place
COD	Chemical Oxygen Demand
CPCB	Central Pollution Control Board
CSR	Corporate Social Responsibility under Companies Act 2013
CW	Cooling Water
CWL	Constructed Wetlands
DBO	Design Build Operate
DO	Dissolved Oxygen
DPR	Detailed Project Report
DT	Disinfection cum Filter Feed Tank
EC	Electrocoagulation
EPA	The Environment (Protection) Act
ERR	Economic Rate of Return
ETP	Effluent Treatment Plant
FINISH	Financial Inclusion Improves Sanitation and Health (Project)
FSM	Faecal Sludge Management
FSTP	Faecal Sludge Treatment Plant
GAC	Granular activated carbon
HCWL	Horizontal Constructed Wetland
HDPE	High density polyethene
HLR	Hydraulic loading rate
HRT	Hydraulic retention time
iMBR	Membrane immersed in the aeration tank
IMT	Intermediate Tank
INR	Indian Rupees
KLD	Kilolitres Per Day
LOMWATS	Low-maintenance Wastewater Treatment System
LW	Liquid Waste
MBBR	Moving Bed Bio Reactor
MBR	Membrane Bio Reactor
MDMS	Ministry of Drinking Water and Sanitation

MF	Microfiltration
MLD	Million Litres per Day
MLSS	Mixed Liquor Suspended Solids
MLVSS	Mixed Liquor Volatile Suspended Solids
MMF	Multimedia filtration
MoEF&CC	Ministry of Environment, Forests and Climate Change
MoUD	Ministry of Urban Development
NF	Nanofiltration
NFT	Nutrient Film Technology
NGO	Non-governmental Organisation
NPO	Nonprofit Organisation
O&M	Operation and Maintenance
Opex	Operating Expenditure
PDU	Pilot Cum Demonstration Unit
PERT	Project Evaluation and Review Technique
PPP	Public–Private Partnership
PSF	Pressure Sand Filter
PST	Primary Settling Tank
RDF	Refuse Derived Fuel
RO	Reverse osmosis
SAFF	Submerged Aerobic Fixed Film
SDB	Sludge drying beds
SLF	Sanitary Landfill
SLWM	Solid and Liquid Waste Management
sMBR	Membrane placed outside of the aeration tank as a side-stream unit
SPBR	Submerged Packed Bed Reactor
SST	Secondary Sedimentation Tank
ST	Settling Tank
STP	Sewage Treatment Plant
SVI	Sludge Volume Index
SW	Solid Waste
SWM	Solid Waste Management
TDS	Total Dissolved Solids
TKN	Total Kjeldal Nitrogen
TSS	Total Suspended Solids
TWT	Treated Water Tank
UAF	Up-flow Anaerobic Filters
UC	Uniformity coefficient
UCB	Urban Council of Balangoda
UF	Ultrafiltration
UNFCCC	United Nations Framework Convention on Climate Change
VCWL	Vertical Constructed Wetland
VGF	Viability Gap Funding

1 Introduction

1.1 WHY THIS BOOK?

This book is essentially focused on waste management in small towns in developing countries. It is written mainly in the Indian context but is adaptable to other developing countries. The field examples are based on the experience and experimentations by the authors, who refer to a few other countries too. While there are several books and textbooks available on various types of wastes and their management, there is no single book covering all types of wastes in the context of a small town. This book can be used as supplementary reading material for different textbooks for students and teachers. The book will be quite handy for quick reference for those working in waste management sectors like administrators and non-governmental organisations (NGOs).

1.2 MUNICIPAL WASTE MANAGEMENT

Waste generation is part of any activity involving living beings, especially fauna. Humans are at the top of the food chain, and hence, the waste generation by human activities are higher when compared to others. Since the density of population as well as industrial activity was low, the waste generation by human activity hardly mattered till the 20th century, as the environment acted as an infinite diluent. This situation changed suddenly with the quantum jump in industrial activity and rapid urbanisation. The rapid increase in human population warranted a higher level of agricultural production too which indirectly increased the industrial activity.

Wastes vary in nature, quantity, and quality depending on the location, activity generating the waste and the receiving environment. By extension, the impact on the environment too varies accordingly. While more often than not, waste is just a misplaced resource, minimisation of the generation and utilisation of the same often present themselves as a complex issue. The waste management hierarchy is the basis of an integrated approach to waste management. It is worth recalling the old puzzle in which an industrial hygiene issue gets converted to a solid waste management problem. The fumes generated from a factory were affecting the workers. In order to tackle this, the management decides to have an exhaust in the production plant. But this became a nuisance to the general public in the form of air pollution. In order to avoid air pollution, a scrubber was installed. The air pollution problem then got converted in to a water pollution problem. Then a wastewater treatment plant was installed by the factory which ended up with some solid waste, which they did not know what to do with!

There is a growing realisation that human activity is having a large deleterious impact on the world's environment, which in turn would affect the survival of humans themselves over a period of time. This realisation and the activities resulting from them are not uniform across the world. Environmental management or waste

DOI: 10.1201/9781003650737-1

management might have a higher priority in developed communities where the people need not be more concerned about food, shelter and secure income. However, we often see developed countries passing on the responsibility to poorer countries in lieu of financial support. India is going through a phase of rapid growth and so now it is in the global frontline of waste crisis. The problem is going to aggravate unless proper measures are taken. Increased level of urbanisation and the population growth are resulting in higher demand for goods and services. One of the direct results is increasing quantities of waste from urban and peri-urban areas. Local self-governments (LSGs) like panchayats, municipalities, and corporations are struggling to find appropriate means of treatment and disposal of these accumulated wastes though the truth is that waste is just misplaced resource. Delhi's Ghazipur landfill is nearly 14 million tonnes of garbage mountain which is nearly 70 m high and similarly many cities are having mountains of garbage, generally termed legacy wastes. Even small towns have their share of legacy wastes which they have to dispose of. What is the way forward to address this? Added to this, there is an issue of faecal sludge/septage management required in the context of absence of adequate ex situ sanitation facilities. In the case of urbanisation, facilities for ex situ sanitation by means of sewage treatment plants are needed. Grey water management is an issue which got obfuscated due to the intense visibility of mountains of solid waste accumulated across the country. However, grey water management is a serious issue requiring much attention since fresh water is becoming dearer by the day and treated grey water can be used for many non-contact applications which in turn reduces the gross fresh water requirements. Massive fish kills in lakes and rivers are being reported very frequently and the main culprit is discharge of untreated sewage into waterbodies which in turn reduces the dissolved oxygen for fish to survive. There are civic bodies which address the issue correctly and make money from waste. There are innovative waste management practices where the civic bodies are earning a substantial amount annually by treating sewage and selling the treated water to industries. Some civic bodies are also converting the organic waste to high value compost and biofuels too. The city's sustainability efforts are enhanced by these methods. Such practices are worth emulating by other civic bodies big and small. There is plenty of opportunities for entrepreneurship in the field of waste management. Success stories of segregating different types of plastics and producing granules, ingots and recycling to make value-added products where virgin plastics are not needed, using recycled plastics for road construction are plenty. Producing quality compost and exporting, producing compost from weeds found in the lakes, and thus restoring the lake and making money out of waste are done by many.

1.3 HOW IS WASTE DEFINED?

Waste (of a material, substance, or by-product) is considered to be 'what has been eliminated or discarded as no longer useful or required after the completion of a process', according to the Oxford Dictionary definition. Waste can be broadly and simply categorised as per its physical properties into solid and liquid waste: if the waste is solid, the waste is called solid waste – as in discarded plastic, paper, glass or metallic objects from residential, industrial and commercial activities in a given area – and if it is liquid form, it is called liquid waste – as in dirty water from houses or chemical discharge from factories which is also called effluent. Mahatma Gandhi called waste a 'misplaced resource', meaning that waste is defined by its 'placing' or 'misplacing' (Anand, 2016). Thus, a slice of bread placed on a plate is food, but the same thing on the ground is waste.

Solid waste is also defined more technically as *discarded material which is in solid form at room temperature and is not easily melted or converted to liquids.* Solid waste is defined by the Government of India (in the Municipal Solid Waste (MSW) Rules of 2016) to mean and include:

> Solid or semi-solid domestic waste, sanitary waste, commercial waste, institutional waste, catering and market waste and other non- residential wastes, street sweepings, silt removed or collected from the surface drains, horticulture waste, agriculture and dairy waste, treated bio-medical waste excluding liquid waste, bio-medical waste and e-waste, battery waste, [and] radio-active waste generated.

A balance between economic development and ecological management is to be the focus when discussing the strategies to be adopted for a viable programme on waste management. If we take the case of food waste alone which is a major constituent of municipal solid waste, studies show that people throw away a third of all the food uneaten, enough to feed the hungry people in the world multiple times. Food waste alone accounts for nearly 10% of the CO_2 emissions (UNFCCC, 2024).

1.4 WASTE MANAGEMENT

It is appropriate to remember that one person's trash is another person's treasure.

A broad perspective that can be adopted for waste management involves the following steps (FS et al., 2017) (Figure 1.1):

1. Avoidance
2. Minimisation
3. Recycling
4. Making value-added products
4. Treatment and reuse
5. Disposal

1.5 APPROACHES TO WASTE MANAGEMENT IN PERI-URBAN AREAS – INTEGRATED WASTE MANAGEMENT

The universally accepted objectives of waste management include

- Achieve total cleanliness in the town
- Make neatness and cleanliness a way of life
- Perceive waste as a misplaced resource
- Generate employment from waste management
- Utilise the resource as a means of revenue

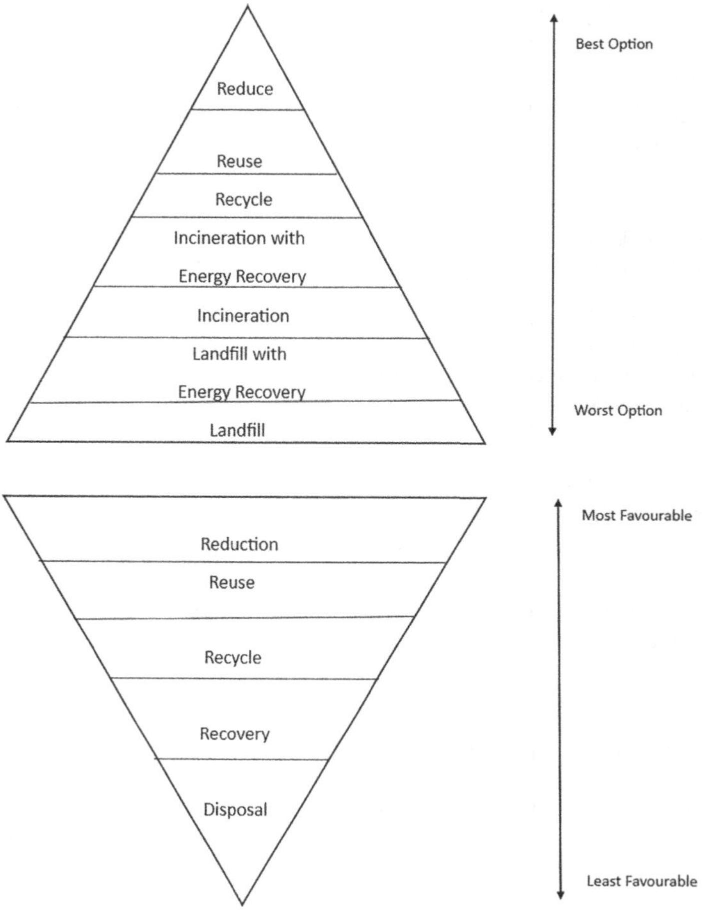

FIGURE 1.1 Waste management hierarchy

- Approach waste management as a basket where cross subsidies within the system can partially sustain it from the financial angle
- Promote entrepreneurship from waste management

1.6 ZERO WASTE ZERO COST CONCEPT

Since we live in a technology supermarket nowadays, the issue in waste management is not technology but management. The primary approach to waste management is to see the resource value of waste, the misplaced resource. The next step is to optimise the waste management processes in such a way that economic viability of the system proposed is ensured along with technical viability. This concept has been extrapolated to the concept of Zero Waste Zero Cost towns. This has some ideological bases too which are given below.

1. Waste is misplaced resource (universal truth but something which cannot be overemphasised)
2. Any problem comes with a built-in solution and the solution itself is the raison d'etre for the problem. The human skill required in solving a problem lies in identifying the solution from within the problem.
3. In any project, the Capital Expenditure (Capex) can be met with by planning without which it will never take off but even with best of planning, meeting the Operating Expenditure (Opex) would pose serious challenges. This applies to waste management too. Generally, waste management is operated on grant-based models but that concept itself undermines the long-term sustainability. Hence, in order to ensure long-term sustainability, the waste management operations should focus on revenue generation. The waste management model should place the whole waste as a basket where some part would generate revenue while the balance may be dealt with as revenue expenditure. The second part effectively gets cross-subsidised by the revenue generation in the first part thereby achieving a Zero Waste Zero Cost situation.
4. Any pollution control measure which uses natural resources has a reasonably significant environmental impact elsewhere. Hence, dependence on natural and finite resources elsewhere needs to be minimised in any method of pollution control to minimise the ecological footprint of the means of pollution control itself.

1.7 CLASSIFICATION OF WASTE

The wastes generated in a small town can be classified as

- organic solid wastes comprising of food wastes, cardboard, paper, leather, plastics, rubber, textiles, wood, yard wastes, agricultural waste etc.,
- inorganic solid wastes, consisting of glass, tin, ferrous and nonferrous metals, ash etc.,
- sewage,
- grey water (sullage),
- septage/faecal sludge,
- medical waste,
- biomedical waste,
- electronic waste, and
- hazardous waste.

The first five can be handled by the local body itself while for the rest, specialist agencies are required. In India, the ideal method to handle them, is to collect and store as per the guidelines of the Central Pollution Control Board (CPCB) and dispose of them to agencies approved by CPCB. *Spurious solutions should not be attempted.* The only material which the local body can handle among these is the diaper/sanitary napkin which, if collected separately, can be disposed of by incineration at a very high temperature in incinerators designed and manufactured/constructed as per the specifications of CPCB. In case there is a large city nearby with a waste to energy (WTE) plant, the diapers and sanitary napkins can be transferred to that plant since they have a very high calorific value and scientific disposal can be ensured.

The peri-urban areas and rural areas have some differences in infrastructure and facilities which influence the way wastes are managed. A peri-urban area will have easy access to the urban markets where certain things can be sold, and some infrastructure for disposal of some of the wastes would be available. Segregated non-degradable wastes can easily be sold in the central market, where we find dedicated shops accepting materials such as metals, segregated or unsegregated plastic wastes etc. There could be potential for manufacturing products from various types of waste which can have a ready market in the city. It can be seen that the key is segregation and value addition. A city would generally be a major consumer of fresh water either from groundwater or from a water body nearby or both. Although a consumer, cities generate sewage which is treated in centralised sewage treatment plants. The grey water generated from a city is not fully directed to the sewage treatment and an appreciable part drains out to some water body through drains and nalas. Another waste generated would be septage/faecal sludge (FS) if the locality has only in situ sanitation facilities. If a centralised Sewage Treatment Plant is available, the FS can be disposed of at the existing centralised Sewage Treatment Plant either directly or through the manholes available in the city. There are cities where dedicated points of disposal of FS are marked and accessible to the operators of Honey Sucker Trucks.

1.8 THE WAY FORWARD

- Identification and classification of wastes for ideal method of reuse/value addition to each component in the waste.
- Develop problem-solving skills for the stakeholders rather than applying excessive focus on technologies. The current world is a technology supermarket. Any technology in the market has to be a good technology; otherwise, it would have gone out of the market long ago. Then, how to select a technology? That is by identifying the most appropriate one for the given context. Contextual solutions can be arrived at by having a viable and easy-to-use decision-making tree.
- Cut down grant-based models and opt for viability gap funding in exceptional cases. Develop a market for any saleable output from the waste management processes. Searching for market, product packaging, pricing, etc., hold the key just like in any commercial venture. Use as many open-source technologies as possible for waste processing.
- Maximise the use of nature-based systems and minimise the dependence of machines needing fossil fuels/electricity for operation; keep the use of machines with moving parts to a minimum to keep control over Opex. Use labour-friendly equipment to keep the labour cost too to a minimum.

1.9 INVOLVING THE SOCIETY IN WASTE MANAGEMENT

Involving the society has several components to it. The authorities are to initiate

- Educating the society on segregation of solid wastes at source
- Incentivise segregation in commercial establishments
- Utilising public facilities for disposal of solid wastes at appropriate places
- Responsible tourism
- Catching them young: Create awareness among school students and incentivise positive action
- Facilitate entrepreneur development and employment generation
- Enactment of strict rules and adherence to the procedures
- Severe penalisation wherever required

1.10 CASE STUDY OF WASTE MANAGEMENT OF A SMALL TOWN IN SRI LANKA

Balangoda is a small town in Ratnapura District in Sabaragamuwa Province in Sri Lanka. The sustainable system developed here is so good that other Municipal councils are also replicating it. The town had a population of 23220 and the area was 16.2 km^2. The town is flanked by two rivers, namely, Walawe and Ve ganga. It has a stream, Dorawela Oya, running through the town too. The Balangoda approach for waste management essentially was focused on techno-commercial viability and dependence on open-source technologies. It was a classic case of circular economy and results included conversion of waste management activities as a source of net revenue income to the town. The Balangoda approach took waste management to the next generation too by involving schools. Bringing the school children into waste management has a direct impact on both the previous and next generations. The approach opened up new avenues for willing entrepreneurs too. The whole process was integrated into one with public participation like motivation in educational institutions, residents' groups, and associations.

There are about 20 tonnes of municipal solid waste generated per day in the jurisdiction of the Urban Council. The main source of waste generation is households, markets, hotels, restaurants, commercial establishments, public and private offices and educational institutions. Street sweeping waste is also combined with the waste management system.

1.10.1 Segregation of Waste at Source and Collection System

The solid waste is segregated into biodegradable and non-biodegradable fractions at the source itself, irrespective of the source. For this purpose, two-bin system is being promoted by the Council. Citizens place the segregated streams of waste by the roads at specified locations and time allotted by the Council authorities. The bio-waste is systematically collected on a daily basis in two shifts, starting from 5 am to 10 pm. Collection of street sweepings from each street is done once in a day and from the main streets twice in a day. For the collection of recyclable materials and other non-biodegradable materials, a separate vehicle is used twice a week. The waste taken to the treatment plant is subjected to secondary segregation. The salient features of waste management process are described here as a case study.

- **Two-way taxation**: This was a unique concept in which the Urban Council taxed commercial units for accepting mixed waste from them but they received a payment for segregated waste. This was to encourage segregation of wastes as reusable, recyclable, non-reusable, non-degradable wastes and degradable wastes. When mixed wastes are received by the Urban Council, the work of segregation falls on it and hence a tax was imposed on the commercial enterprises. The taxes too varied according to the nature of the enterprises like hotels, restaurants, shops, and small food industries. There are small establishments like bakeries and teashops which were finding segregation of solids difficult, and they were offering only unsegregated solid wastes to the Urban Council. In order to reduce/control this tendency, a proposal for two-way taxation was mooted and implemented. The system offered a fixed rate for segregated wastes, but the Urban Council charged the establishments for unsegregated wastes. For this purpose, the establishments were categorised and taxed with different rates.
- **Secondary segregation**: The segregated biodegradable waste brought to the waste treatment plant is again segregated for any non-biodegradable material that is present in it by feeding to a slowly moving (5 metres/minute) conveyor belt and the non-decomposable material such as plastics, glass, and metals are manually removed by workers standing on either side of the conveyor belt. The workers are provided with hand gloves and they manually remove the material from the moving belt (the thickness over the belt is kept less than 150 mm) and the removed material is stored separately.
- **Mechanical Size Reduction**: The waste is then subjected to size reduction where the surface area per unit weight is increased for faster biological decomposition. Size reduction also helps in reducing fly breeding in the decomposing mass. This is commonly carried out either in hammer mills or rasp mills. Hammer mills are high-speed (600–1200 revolutions per minute) compact machines, but consume large amounts of energy. Rasp mills are slow-moving large units that require lesser energy. The capital cost of a hammer mill is less, but its operating cost is more than that of a rasp mill, mainly due to the larger energy requirement as well as more frequent replacement or re-tipping of hammers.
- **Processing of biodegradable waste**: The Council operates a 12 tonnes per day (TPD) windrow compost plant at Bankiyawatha in Balangoda. The capacity of the plant is enough to manage all the waste collected, since the quantity of the remaining waste after home composting, use of food and vegetable waste in poultry farms and pig farms and utilisation in agriculture, is practically less than 12 tonnes. All the wet waste collected daily is transported to the plant and subjected to aerobic composting after secondary sorting to remove non-compostable materials.

- **Stabilisation in Open Windrows**: The stabilisation is carried out in open windrows provided over flagstone-paved or cement-concrete-paved ground. These windrows are turned every five days to ensure aerobic decomposition. Various types of equipment such as front-end loaders/windrows shifters are used for turning of windrows. At the end of the three to four weeks period, the material is known as green or fresh compost, wherein the cellulose has not been fully stabilised. It is hence stored and cured in large sized windrows for one to two months at either the plant or the farms. At the end of the curing period, it is known as ripe compost. It may be sometimes subjected to size reduction to suit kitchen garden and horticulture requirements.
- **Dedicated Faecal Sludge Treatment Plant (FSTP)**: FSTP was set up to treat the Faecal Sludge (FS) collected from Septic Tanks and Leach Pits in the town. The collection and transportation of FS was managed exclusively by honey sucker equipment and trucks owned and operated by the Urban Council.
- **Super compost**: Co-compost with FS trials with different ratios of treated FS and Bio Compost were made to arrive at an optimum ratio of mixing. A particular combination of the mix was found to be superior. It was sold as Super Compost. Phase 2 was using Bio Compost and Treated Septage as a platform for preparing a superior product.
- **Involvement of an R&D organisation**: Tea Research Institute of Sri Lanka was involved for the assessment of the products and suggestions for improvements. After several trials, a combination consisting of treated faecal sludge, biocompost, powdered rock phosphate, and biochar was found to be an ideal manure for many crops.

  As Phase 3, the new product was tried for various crops and optimum quantities of application were arrived at. The new product called fortified compost started receiving acceptance in different agricultural sectors.
- **Segregation, classification, and processing of non-degradable solid wastes**: Segregation of non-degradable wastes has been tried at different levels and up to 120 types of items had been identified in a typical urban case. However, such type of segregation would be too cumbersome and counterproductive in small towns. Hence, a more practical level of classification and segregation was adopted. The strategy was to identify components which could be sold directly and components which could be processed to achieve significant value addition. The key is the identification of materials which can be isolated and processed to achieve value addition. Additional revenue generation is achieved by segregation of saleable items.
- **Entrepreneur development by setting up Sampath Kendras**: Sampath Kendras was a multi-dimensional concept. Waste collection centres were set up in various parts of the town. In order to attract the public, they were named Sampath Kendra, which means Wealth Centre. The entrepreneur was free to buy waste from anyone and segregate it further and sell each of the segregated items in the appropriate market, generally located in Colombo and Kandy. The prices for each item would be fixed by the Urban Council and displayed in the Sampath Kendra, and often revised. The Urban Council released dedicated buildings to entrepreneurs interested to run Sampath Kendras. These

buildings are located in prominent junctions to facilitate easy access for the citizens. They were given temporary financial assistance by the Urban Council to commence the operations. There would be a list of materials accepted in the Sampath Kendra with the prices displayed there. Such materials could be sold to the Sampath Kendras by any citizen. The entrepreneur would be linked to the recycling units in cities like Kandy and Colombo to whom the entrepreneur can sell the items and keep the proceeds himself/herself. Thus, the Sampath Kendras ensure the cleanliness of the town, provide business opportunities, and create jobs for those who work in the Sampath Kendras.

- **Motivation at schools**: Sampath Kendras were set up within schools too. The data from the Sampath Kendras indicated that most of the solid wastes were brought there by school children. The students were provided with cards in which the daily value of the solid wastes received was marked. At the end of each month, the students were paid in cash according to the recordings on the card. Further, the students collecting the highest quantity of waste were given special recognition in the form of rewards, badges, felicitation in school assemblies, etc.
- **Critical component identified in value addition**: From the non-degradable materials in Balangoda, the high-density polyethylene (HDPE) component of the plastic wastes was separated and processed to produce granules which had a sale value of seven to eight times that of the mixed plastic wastes. A separate plant was set up for cleaning and making granules of HDPE. The operational costs were met with the sale proceeds and there were enough funds left with to cross-subsidise sanitary landfill and similar activities without cash inflow.
- **Low Maintenance Wastewater Treatment systems**: Going beyond Sanitation and Waste Management, the Urban Council undertook installation of simple Low Maintenance Wastewater Treatment systems for various

FIGURE 1.2 Compost piles under preparation. Photo taken by the author

drains joining the two rivers and undertook the rehabilitation of Dorwela Oya. The drains essentially discharging grey water into the rivers did receive some fugitive effluents too. The inline treatment systems ensured better quality of the water drained into the rivers. The Urban Council (UC) undertook

FIGURE 1.3 Compost product ready for bagging. Photo taken by the author

FIGURE 1.4 Super compost bags indicating the quantity to be applied for each crop. Photo taken by the author

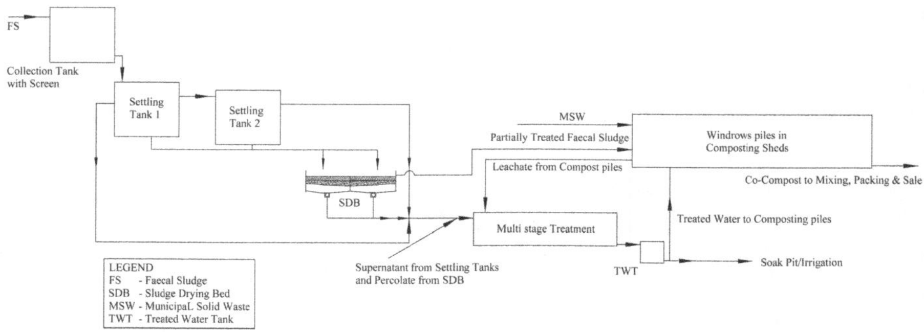

FIGURE 1.5 Process flow diagram of co compost plant

rehabilitation of Dorwela Oya which had some international support too. The Dorawela Oya ensured adequate groundwater in the town and its pollution was affecting the groundwater sources. Accumulation of debris on its bed had been a cause for floods in parts of the town. The cleaning of the stream was taken up with community involvement. This action had multiple results like minimising of floods in the town during rainy seasons, improving the quality of groundwater, and providing a recreational space within the town. Planting of trees along the stream too was taken up. In some portions, Vetiver was planted both as a line of protection and a means of soil stabilisation.

- **Sustainability and Replication of the Balangoda system**: The Urban Council of Balangoda ended the financial years with positive net revenue from waste management for the first five years after stabilisation of the waste management system. The net revenue income indicated sustainability of the system. The inputs to achieve this included commitment, technology, institutionalisation of the systems and political will. The sustainability of the system has triggered replication/emulation in several small towns in Sri Lanka.

1.11 CLIMATE IMPACT AND ECOLOGICAL IMPACT

Waste Management became a necessity due to generation of waste. Any action to treat waste of any kind, solid, liquid, gaseous, etc., has an ecological footprint and a climate impact. However, no action in waste management would have far worse climate impacts. For example, solid wastes left to rot on their own have a far worse impact than any treatment the solid wastes undergo. The selection of technologies is basically dependent on various factors like

- Availability of enough land in a proper locality
- Availability of power
- Availability and need for skilled manpower
- Availability of equipment
- Capital Expenditure (Capex) and Operating Expenditure (Opex)

- Ecological impact
- Climate impact
- Long-term techno-economic viability
- Implications of Intellectual Property Rights (IPR)

In order to limit Opex, the best way forward would be open-source technologies with greenest of approaches. Nature-based solutions top the list. Wherever use of green energy is possible, they come next. It may be of interest to note that even the greenest of energy need not necessarily be of zero impact on ecology or climate since there would be materials of construction which would have had some impact elsewhere. For example, if wind energy is used, the windmill needs to be manufactured using finite resources elsewhere and in a solar-based power generation, the materials used in the batteries and panels are from finite sources.

Next comes the comparison of technologies. When composting of degradable solid wastes is considered, aerobic windrows take the top slot as practically no machinery is used and minimum generation of greenhouse gases. When it comes to treatment of liquid wastes like faecal sludge, black water or grey water, nature-based systems like constructed wetlands take the top slot.

1.12 LONG-TERM SUSTAINABILITY

The long-term sustainability of a system depends on various factors, of which political commitment and administrative continuity are the top-most requirement. There had been excellent, economically independent, and self-sustaining models which were affected by administrative apathy or lack of interest of the political leadership. Although self-sustaining systems need not be individually dependent per se, the administrative apathy when there is a change of officials take place do seriously impact the performance of the system. Selection of technology does have a long-term impact on sustainability. Technological obsolescence is a major issue. Likewise the cost of replacement of components after a certain period of operations need serious Capex which would not be factored in originally. The change in the structure of the society too will have a significant impact since availability of appropriate type of manpower too would be a concern with respect to long-term sustainability.

1.13 CLIMATE CHANGE AND SOLID WASTE MANAGEMENT

Climate change is the result of global warming. Global warming refers to the long-term rise in the earth's surface temperature due to the increasing levels of greenhouse gases in the atmosphere, namely, carbon dioxide, methane, and oxides of nitrogen. Anthropogenic activities release greenhouse gases in excess of naturally occurring amounts which become the main source of climate change by adding carbon-based molecules to the atmosphere. Climate change is expected to increase earth's mean temperature and sea level rise, increasing frequency and severeness of droughts, floods, heat waves, and water scarcity. Polar ice caps melt, marine ecosystems get destroyed, leading to increased food shortages and migration. Plants and animal life suffer due to temperature rise. Solid waste is identified as one of the sources of greenhouse gas emissions. It is estimated as contributing about 5% of global greenhouse gas emissions in the form of carbon dioxide, methane, and nitrogen oxides. The most significant greenhouse gas is the methane gas produced from landfill which is mostly released from the breakdown of organic matter in the absence of oxygen from anaerobic digestion. Meanwhile inorganic wastes do not contribute directly to greenhouse gas emissions unless it is incinerated or burnt. The solid waste processing methods and systems discussed in this book are supposed to address the carbon emission issue in a big way. Both the biodegradable waste processing systems and non-biodegradable systems when managed scientifically will reduce the carbon emission considerably. International and national efforts towards climate-friendly and sustainable waste management should follow the waste management hierarchy. It prioritises waste prevention, waste reduction, recycling (composting and anaerobic digestion for biodegradable and recycling for non-biodegradables), and energy recovery from waste before land filling or burning.

2 Legislation for Environmental Protection

2.1 REGULATORY ASPECTS IN THE GLOBAL SCENE

Rapid industrialisation and consequent urbanisation led to increasing pollution of the environment. Population growth and the need for chemical fertilisers and chemical pesticides for food production to feed the masses were essential. But the indiscriminate use of pesticides because of lack of awareness of the side effects and the harm caused to the environment was realised only later. Realisation of this as a global issue and subsequent enactment of laws and regulations became significant only from the 1970s. The book 'Silent Spring' by Rachel Carson brought the topic to the public eye and created awareness as to the impact of human activities on environment. Rapid urbanisation without adequate access to safe sanitation facilities is a major issue faced by all developing countries. The indiscriminate disposal of domestic wastewater is the main reason for degradation of water quality with severe impact on health, environment and thus the economy. The Stockholm Global Conference on Human Environment (1972) triggered public action through legislation and other measures to protect and promote environmental quality around the world. This conference raised global awareness about environmental issues and the need for international participation for sustainable growth. The first United Nations conference inspired subsequent initiatives and was followed by the Montreal Protocol (1987) to combat Ozone depletion, the Rio Declaration (1992) in Rio de Janeiro Earth Summit, etc. The Convention on Climate Change 1995 created the Conference of the Parties (COP) meeting every year, and the Kyoto Protocol (1997) was presented. The first international agreement with legally binding obligations for developed countries (2002) was the Declaration of Sustainable Development in Johannesburg, in which 190 countries attended. The Paris agreement (2015) was adopted to prevent the temperature rise of the planet by 2°C and limit it to 1.5°C, with a focus on bringing it down to the preindustrial period. The Baku Conference from November 11 to 22 in 2024, officially COP 29 was the latest in the series of COPs where nearly 200 countries participated and the key focus was climate finance, mitigation goals, adaptation strategies, and stock taking. All countries started enacting environmental laws and regulations to protect the environment in line with the Stockholm conference from 1972 onwards.

2.2 ENVIRONMENTAL REGULATIONS IN INDIA

The earliest traceable environmental legislation is by Emperor Ashoka in the 3rd century BCE, as seen in his edicts on wildlife protection, afforestation, planting of

DOI: 10.1201/9781003650737-2

medicinal trees and the importance of sustainable living in harmony with nature. Ancient scriptures like the Vedas and Upanishads mention environmental protection for sustainable living. Archaeological evidence of the Indus Valley civilisation which dates back 5000 years shows that hygiene, waste management and sanitation, underground drainage system were all practiced by the people then.

Charaka Samhita, the Ayurveda medical system text (600 BCE), has slokas, which instruct methods of water purification. Notable legislation in the British regime for protecting the environment is the Cattle Trespass Act 1871, the Elephant Preservation Act 1879 and the Indian Forest Act 1927. India enacted the Water (Prevention and Control of Pollution) Act in 1974 and Air (Prevention and Control of Pollution) Act in 1981 in order to have effective regulatory measures at government level to protect environment. While it is the Parliament which legislates in the case of Pollution Control Laws, the State Legislatures have the freedom to amend them to make them stricter but not lenient. The State level Pollution Control Boards (SPCB) are responsible for enforcing the pollution control laws. The primary function of Central Pollution Control Board (CPCB) is to advise the Union Government on pollution control. CPCB has the additional responsibility of research, collection, and publishing of data, and generating guidelines and manuals to support the Union government and State governments as well as the State Pollution Control Boards in developing standards and implementation of pollution control policies. Developing the standards, inspecting potential sources of pollution, monitoring of pollution, enforcing the pollution control laws, including legal actions, etc., fall under the purview of State Pollution Control Boards.

2.3 ENVIRONMENTAL POLICY

The Government of India had initiated many policies related to the protection of environment since 1968. It started with the National Forest Policy of 1968. The Indian Constitution, in the section on Fundamental duties, have stated that 'To protect and improve the natural environment including forests, lakes, rivers, and wildlife, and to have compassion for living creatures' is a fundamental duty {(Part IV A) Article 51A}. Post-independence, the law-making started with wildlife protection but moved into various fields in a step-by-step manner.

2.3.1 The Wildlife (Protection) Act (WPA) 1972

The Act provides for the protection of wild animals, birds, and plants; and for matters connected therewith or ancillary or incidental thereto. It extends to the whole of India. It has six schedules that give varying degrees of protection:

- Schedule I and Part II of Schedule II provide absolute protection, offences under these are prescribed the highest penalties.
- Species listed in Schedule III and Schedule IV are also protected, but the penalties are much lower.
- Animals under Schedule V, e.g. common crows, fruit bats, rats, and mice, are legally considered vermin and may be hunted freely.

- The specified endemic plants in Schedule VI are prohibited from cultivation and planting.

Statutory bodies under WPA:

a. National Board for Wildlife and state wildlife advisory boards
b. Central Zoo Authority
c. Wildlife Crime Control Bureau
d. National Tiger Conservation Authority

The acts and rules for Prevention and Control of Pollution and Conservation of groundwater followed this. The series of acts by the Parliament started with the Water (Prevention and Control of Pollution) Act which came into force in 1974. In order to control and monitor the mining and usage of groundwater, the Water Cess Act came into force in 1977. In 1980, the Government of India established the Department of Environment. Post formation of the Ministry, the Air (Prevention and Control of Pollution) Act was initiated in 1980, and the rules were framed in 1981. Combining the aspects of different pollution control acts and rules, the Environment Protection Act was formed in 1986. The salient features of these acts are summarised below.

2.3.2 The Water (Prevention and Control of Pollution) Act, 1974

Objective: To provide prevention and control of water pollution. Maintaining or restoring of wholesomeness and purity of water in the various sources of water.

Water Pollution in India is regulated under the provision of the above Act. The salient features of the act are

- It vests regulatory authority in Central Pollution Control Board and State Pollution Control Boards.
- Central Pollution Control Board (CPCB) and State Pollution Control Board (SPCB) are statutory bodies created under the Water Act of 1974. It empowers CPCB and SPCB to establish and enforce effluent standards for factories discharging pollutants into water bodies.
- CPCB performs these same functions for union territories along with formulating policies related to the prevention of water pollution and coordinating activities of different SPCBs.
- SPCB controls sewage and industrial effluent discharge by approving, rejecting, and granting consent to discharge.
- The Act generally prohibits the disposal of polluting matters in streams, wells, sewers, or land in excess of the standards established by state boards. A permit system wherein consent is to be obtained from the state board before establishing any industry operation, trade, process or any treatment and disposal system or any addition to such systems which will result in the discharge of sewage or effluent to a stream, well, or land.
- Under this Act there is no difference between industrial and domestic pollutants.

Due to the lack of attention and suitable planning domestic pollution is not properly regulated and hence this is becoming the biggest polluter in India. Government of India launched the Ganga Action plan and National River Action plan and so many Sewage Treatment Plants (STPs). Still the situation is alarming.

2.3.3 The Air (Prevention and Control of Pollution) Act, 1981

The Act aims to control and prevent air pollution in India and its main objectives are:

- To provide for prevention, control, and abatement of air pollution.
- To provide for the establishment of the boards at the central and state levels to implement the act.
- CPCB and SPCB were given the responsibility.
- It states that the sources of air pollution such as internal combustion engines, industry, vehicles, and power plants are not permitted to release particulate matter, lead, carbon monoxide, sulphur dioxide, nitrogen oxide, volatile organic compounds (VOCs), or other toxic substances beyond the predetermined limit.
- It empowers the state government to designate air pollution areas.

2.3.4 The Environment (Protection) Act, EP Act, 1986

This act was passed under Article 253 (legislation for giving effect to international agreements) and was passed in the wake of the Bhopal gas tragedy in December 1984. It was enacted to achieve the goals of UN Conference on the Human Environment, 1972 – Stockholm Declaration. Statutory bodies under the EP Act, 1986:

a. Genetic Engineering Appraisal Committee
b. National Coastal Zone Management Authority
c. National Ganga River Basin Authority
d. National Environmental Appellate Authority (Formed later by a separate Act relevant to EP Act 1986)
e. Central Ground Water Authority constituted under Section 3(3) of EP Act 1986

This Act empowers the Union Government to take all measures as it seems necessary for the purpose of protecting and improving the quality of the environment and preventing, controlling, and abating environmental pollution. It takes into account the entire ecosystem which includes water, air, land, and interrelationship among water, air, land, human beings, other creatures, plants, microorganisms, and property. The act empowers the government to establish standards for the quality of the environment in its various aspects, including maximum allowable concentration of various environmental pollutants including noise for different areas. EP Act was the first environmental statute to give the government authority to issue direct written orders including orders for closure, prohibit, or regulate any industry or process operation. Specific provisions for handling of hazardous material are also included. The act provides for severe penalties for those who do not comply with the provisions or rules and directions issued by the Act. A citizen's suit provision is also included in this Act.

The Act explicitly prohibits the discharge of any pollutant above the limits prescribed in the act and any person who fails to comply will be punished for each failure with fine and prison term up to 5 years or both. To ensure compliance with the Act, it permits power to entry for examination, testing of equipment or any other purpose and taking of samples of air, water, soil, or any other material deemed necessary for analysis. By this Act the Union Government is authorised to set new national standards for ambient quality of the environment and standards for controlling emissions and effluent discharges, to regulate industrial locations, to prescribe procedures for managing hazardous substances, to establish safeguard for preventing accidents and to collect and disseminate information regarding environmental pollution.

This Department of Environment later became a full-time Ministry as the Ministry of Environment and Forests. The Ministry was renamed later in 2014 as the Ministry of Environment, Forests, and Climate Change (MoEFCC). Eco-sensitive zones or ecologically fragile areas are notified by under EP Act, 1986 – 10 km buffer zones around protected areas. Post formation of the Department of Environment, the policies adopted by the Government of India include

a. National Conservation Strategies and Policy on Environment and Development (1992)
b. Policy Statement on Abatement of Pollution (1992)
c. National Environment Action Programme (1994)
d. National Agricultural Policy (2000)
e. National Population Policy (2000)
f. National Water Policy (2002)
g. Wildlife Conservation Strategy (2002)

The National Environment Policy came into force by 2006.

The laws on Pollution Control actually started with the old Public Nuisance Act. There are several sections in Indian Penal Code (IPC) and Criminal Procedure Code (CrPC) for dealing with cases pertaining to pollution, the details of which are given below.

Section 268 – Public Nuisance
Section 269 – Negligent Act likely to spread infection of disease dangerous to life
Section 270 – Malignant Act likely to spread infection of disease dangerous to life
Section 277 – Fouling of water of public spring or reservoir
Section 278 – Making atmosphere noxious to health
Section 284 – Negligent conduct with respect to poisonous substances
Section 285 – Negligent conduct with respect to fire or combustible matter
Section 286 – Negligent conduct with respect to explosive substances
Section 287 – Negligent conduct with respect to machinery
Section 288 – Negligent conduct with respect to pulling down or repairing buildings

Even as late as 2024, *Section 268* and *Section 277* over and above exclusive pollution control laws were slapped by the police on an industry for polluting a river, resulting in a massive fish kill. This incident shows the effectiveness of the old general laws even after forming laws dedicated exclusively for protection of environment. While the general laws indicated above can effectively be used in case of pollution caused deliberately or otherwise, exclusive laws were necessary to deal with specific situations.

2.3.5 The Ozone-Depleting Substances (Regulation and Control) Rules, 2000

- It set deadlines for phasing out various ozone-depleting substances (ODSs) and regulating production, trade import, and export of product containing ODS.
- These rules prohibit the use of chlorofluorocarbons (CFCs), halons, ODSs such as carbon tetrachloride, and methyl chloroform.

2.3.6 Coastal Regulation Zone Notification 2018

It was notified based on the recommendations of the Shailesh Nayak Committee.

- To promote sustainable development while taking into account the natural hazards such as increasing sea levels due to global warming.
- To conserve and protect biodiversity, besides livelihood security for local communities including the fishermen.

Coastal Regulation Zones (CRZs) have been classified into four zones for regulation:

i. CRZ I – ecologically sensitive areas such as mangroves, coral reefs, salt marshes, turtle nesting ground, and the inter-tidal zone.
ii. CRZ II – areas close to the shoreline, and which have been developed.
iii. CRZ III – Coastal areas that are not substantially built up, including rural coastal areas.
iv. CRZ IV – water area from low tide line (LTL) to the limit of territorial waters of India.

2.3.7 The Energy Conservation Act, 2001

It was enacted as a step towards improving energy efficiency and reducing wastage. It specifies the energy consumption standards for equipment and appliances. It prescribes energy consumption norms and standards for consumers. It prescribes energy conservation building codes for commercial buildings. The Bureau of Energy Efficiency (BEE) is a statutory body established under the act.

2.3.8 Biological Diversity Act 2002

It was implemented to give effect to the Convention on Biological Diversity (CBD), Nagoya Protocol.

- To check biopiracy, protect biological diversity, and local growers through a three-tier structure of central and state boards and local committees.
- To set up the National Biodiversity Authority (NBA), State Biodiversity Boards (SBBs), and Biodiversity Management Committees (BMCs).

2.3.9 Scheduled Tribes and Other Traditional Forest Dwellers (Recognition of Forest Rights) Act, 2006 (FRA)

The act recognises and vests the forest rights and occupation in forest land in Forest Dwelling Scheduled Tribes (FDST) and Other Traditional Forest Dwellers (OTFD) residing in such forests for generations. This act comes under the aegis of the Ministry of Tribal Affairs.

- The act also establishes the responsibilities and authority for sustainable use, conservation of biodiversity, and maintenance of the ecological balance of FDST and OTFD.
- It strengthens the conservation regime of the forests while ensuring the livelihood and food security of the FDST and OTFD.
- It seeks to rectify colonial injustice to the FDST and OTFD, who are integral to the very survival and sustainability of the forest ecosystem.
- The act identifies four types of rights

i. Title rights: It gives FDST and OTFD the right to ownership of land farmed by tribals or forest dwellers subject to a maximum of 4 hectares. Ownership is only for land that is being cultivated by the concerned family, and no new lands will be granted.
ii. Use rights: The rights of the dwellers extend to extracting Minor Forest Produce, grazing areas, pastoralist routes, etc.
iii. Relief and development rights: Right to rehabilitation in case of illegal eviction or forced displacement and basic amenities, subject to restrictions for forest protection.
iv. Forest management rights: It includes the right to protect, regenerate, conserve, or manage any community forest resource that they have been traditionally protecting and conserving for sustainable use.

2.3.10 The National Green Tribunal (NGT), 2010

It was established in concurrence with the Rio Summit 1992 to provide judicial and administrative remedies for the victims of the pollutants and other environmental damage. It also agrees with Article 21, the Right to a healthy environment for its citizens, of the constitution. The NGT has to dispose of the cases presented to it within 6 months of their appeals. NGT has original jurisdiction on matters related to substantial questions of the environment.

NGT deals with the civil cases under the seven acts related to the environment:

a. Water (Prevention and Control of Pollution) Act, 1974
b. Water (Prevention and Control of Pollution) Cess Act, 1974

c. Air (Prevention and Control of Pollution) Act, 1977
d. Forest Conservation Act, 1980
e. Environmental Protection Act, 1986
f. Public Liability Insurance Act, 1991
g. Biological Diversity Act, 2002

Two acts have been kept out of the jurisdiction of NGT:

i. Wildlife Protection Act, 1972
ii. Scheduled Tribes and Other Traditional Forest Dwellers (Recognition of Forest Rights) Act, 2006 (FRA)

The decisions of the NGT can be challenged in High Courts and the Supreme Court.

2.3.11 Compensatory Afforestation Fund (CAF) Act, 2016

The CAF Act was enacted to manage the funds collected for compensatory afforestation which till then was managed by ad hoc Compensatory Afforestation Fund Management and Planning Authority (CAMPA). Compensatory afforestation means that every time forest land is diverted for non-forest purposes such as mining or industry, the user agency pays for planting forests over an equal area of non-forest land, or when such land is not available, twice the area of degraded forest land. As per the rules, 90% of the CAF money is to be given to the states while 10% is to be retained by the Centre. The funds can be used for the treatment of catchment areas, assisted natural generation, forest management, wildlife protection and management, relocation, etc.

The National Faecal Sludge & Septage policy was developed in 2022. The pollutants include general municipal solid wastes, biomedical wastes, and hazardous wastes too. Efforts to address the issues resulted in framing rules for managing these wastes. Sanitation had been a big issue in India for quite long. The issue of sanitation was initially addressed by the Total Sanitation Campaign (TSC) whose successor programmes were Nirmal Bharat Abhiyan (NBA) and Swachh Bharat Mission (SBM). Unlike industrial wastes which can easily be controlled due to specific locations of sources of pollutants and pollution, managing pollution from household liquid wastes, both sewage and sullage is a big challenge. In the case of ex situ sanitation, there would be a sewer system transporting the liquid wastes to a centralised Sewage (Wastewater) Treatment Plant and monitoring the performance of the plant would ensure minimisation of pollution. However, if it is case of in situ sanitation, the management of faecal sludge and grey water would require extensive network for collection, transportation, and treatment over and above having in situ solutions. While in situ sanitation and management of faecal sludge/septage has been extensively covered in the Central Public Health and Environmental Engineering Organisation (CPHEEO) Manual, there is very little work done in the direction of Grey Water Management.

A list of other important legislation is listed in Table 2.1.

TABLE 2.1
List of Legal Provisions

Central/State Legal Provisions	Main Objectives
Policy Statement for Abatement of Pollution, 1992	Suggests developing relevant legislation and regulation, fiscal incentives, educational programmes, information campaigns, pollution prevention at source, focus on heavily polluted areas and river stretches and involve public in decision making
National Environment Policy, 2006	Promotes conservation of natural resources, protection of wildlife and ecosystems, prevention of pollution, reuse and recycling of waste water, adoption of clean technology, application of polluter pays principle, etc.
National Sanitation Policy, 2008	Awareness generation, behaviour change, open defecation free cities, safe disposal of wastes and proper operation and maintenance of sanitary installations. The different states and cities have to develop their sanitation policies, strategies, and goals
The Public Liability Insurance Act, 1991	To provide for public liability. Insurance for the purpose of providing immediate relief to the persons affected by accident occurring while handling any hazardous material
Hazardous Waste (Management and Handling), Rules, 1989 as Amended in 2016	Legislative framework for the law enactment related to storage, transport, handling, and disposal of hazardous wastes
Manufacture, Storage and Import of Hazardous Chemical Rules, 1989	Rules for manufacture, storage, and import of hazardous chemicals
Municipal Solid Waste (Management and Handling) Rules, 2000 Amended as Solid Waste Management Rules 2016	Rules for Municipal Solid Waste Management of Urban Cities Applicability expanded to every urban local body, outgrowths in urban agglomerations, census towns, notified areas, notified industrial townships, areas under the control of Indian Railways, Airports, Air Bases, Ports & Harbours, Defence Establishments, Special Economic Zones, State and Central Organizations, Places of Pilgrims, Religious & Historical importance, Every domestic, institutional, commercial, and any other non-residential solid waste generator situated in the areas except Industrial Waste, Hazardous Chemicals, Hazardous Chemicals, BMW, E Waste, Lead Acid batteries and Radio Active waste, that are covered under separate rules framed under the EP Act 1986…
The National Environment Tribunal Act, 1995 Replaced with NGT Act 2010	To provide for strict liability for damages arising out of any accident that results in death or injury of any person or if it causes any harm to the environment
Environment Impact Assessment Notification, 2006	To grant green clearance for establishment or expansion on an industry on the basis of the potential environmental impact of the project
Municipalities Act, District Municipalities Act, or the Nagar Palika Act	Complete Authority and jurisdiction over all urban amenities, including Waste Management, water supply and sanitation with the municipality

Source: Rajagopalan (2015), Naresh (2020).

3 Municipal Solid Waste Management

3.1 INTRODUCTION

Municipal solid waste (MSW) is generated by human activities and may contain several types of wastes including biodegradable food wastes, agricultural wastes, non-biodegradable wastes like paper, glass, cardboard, cans, leather, clothes, and rubber products. MSW may also contain medical, biomedical, and hazardous wastes too. If left untreated, MSW leads to degradation of the environment. Degradable materials in MSW decompose to generate foul-smelling and easily combustible gases and may cause fire hazards. To ensure a safe disposal of the MSW, it is essential to minimise environmental pollution and fire hazards. There are several processing methods available for this purpose. Odour develops in onsite storage like collection centres, transfer stations, and dumping grounds and in warm climate it is even more significant. The breeding of flies is another important problem for organic waste storage in warm climates. Many of the non-degradable wastes are useful as either raw materials or as saleable products. Segregation ensures no mixing of such components with other items. Mixing reduces their value and increases the complexities of processing. There would be hazardous components which need to be segregated and processed separately by appropriate means of disposal. The same approach would be needed while handling medical and biomedical wastes. Though there are many methods available for processing and handling of MSW, quite often local bodies like Panchayats and Municipalities of small towns in developing countries are finding it difficult to process the waste appropriately. The general characteristics of MSW and the standard methods available for processing are described in detail here. Case studies of successful models implemented are described.

3.2 CHARACTERISTICS OF MSW

3.2.1 Properties of MSW

The physical, chemical, and biological properties of the municipal solid waste (MSW) must be known for dealing with the waste for appropriate management. The nature and characteristics of the waste determine the selection of processes for an integrated waste management system.

3.2.2 Physical Properties

Most important *physical* properties are density, moisture content, field capacity, compacted waste porosity, particle size, and particle size distribution.

 DOI: 10.1201/9781003650737-3

3.2.2.1 Density

Density of a waste is the mass per unit volume expressed as kg/m^3. It is essential for the design of all elements of the solid waste management system from storage to transportation to final disposal. There will be significant changes in density as the waste is taken from source to disposal site due to scavenging, handling, wetting, and drying due to weather conditions. Density varies from 350 to 550 kg/m^3 in some cities in India.

3.2.2.2 Moisture Content

Moisture can be of three types. Surface moisture, absorbed moisture, and chemically bound moisture. Moisture increases the weight of solid waste, and hence, the cost of collection and transportation is increased. It is a major factor for the economic feasibility of treatment by incineration, as wet waste needs energy for evaporation of water. The moisture adhering to the surface of the sample is the surface moisture. The moisture that is present in the pores and capillaries is the absorbed moisture, and the moisture that is chemically bonded to other molecules is the chemically bonded moisture. For MSW, a thermogravimetric method is used for moisture content determination (IS, 1979). The sample is heated and the weight loss due to evaporation is noted. Moisture content can be expressed in dry method or wet method. For MSW, the wet weight method is used. The sample is dried in an oven till its mass becomes constant. The sample is spread thinly in an oven at a temperature of 378 K (105°C). In the case of combustibles, the temperature shall be kept at 343–348 K (70–75°C) and if microbial tests are to be conducted, drying shall be done at 313–323 K. The wet weight (W) and dry weight (D) are noted and the weight loss ($M = W - D$) is calculated and expressed as a percentage of the weight of the wet sample.

For wet basis calculation

$$M = \left(\frac{W - D}{W}\right) \times 100 \tag{3.1}$$

M = Moisture content %
W = Weight of sample, kg
D = Weight of sample after drying, kg

The moisture content of MSW varies from 15 to 45% in general.

Example 3.1

Calculate the moisture content of a certain municipal solid waste with the following composition (Table 3.1).

TABLE 3.1
Municipal Solid Waste Sample Analysis

Type of Waste	Wet Weight (kg)	Dry Weight (kg)
Food waste	30	10
Paper	15	13
Yard waste	25	15
Others	30	17

Solution

Basis: 100 kg wet sample
Moisture content of municipal solid waste = M
Substituting in Equation 3.1

TABLE 3.2
Calculation of Moisture Content for Example 3.1

Type of Waste	W = Wet Weight	D = Dry Weight	$(W - D)$ =Moisture
Food waste	30	10	30 - 10 = 20
Paper	15	13	15 – 13 = 2
Yard waste	25	15	25 – 15 = 10
Others	30	17	30 – 17 = 13
Total	100	55	45

$$M = \left(\frac{W - D}{W}\right)100$$

$$M = \left(\frac{100 - 55}{100}\right)100 = 45\% \qquad \text{(Answer)}$$

3.2.2.3 Particle Size and Size Distribution

Measurement of size and size distribution are important factors in solid waste management because of their significance in the design of shredders and mechanical separators. The size of a waste component is expressed in many ways, as the waste contains different types of materials like paper, cardboard, plastics, textiles, food waste, and yard waste and so a mean particle size may not be representative. The size is important when we are using trommel screens and magnetic separators. The size s can be represented in the following ways:

$$s = l \quad s = \frac{l + b}{2} \quad s = \frac{l + b + h}{3}$$

$$s = \sqrt{l \times b} \quad s = \sqrt[3]{l \times b \times h} \qquad (3.2)$$

where

s = size of the component in mm
l = length mm
b = width mm
h = height mm

The size and size distribution of waste components are also important for waste recovery. This information helps in the utilisation of mechanical separators and shredders for waste stream processing. The major means of controlling particle size is through shredding. The advantages of shredding are that it increases the homogeneity, increases the surface to volume ratio and reduces the potential for preferential liquid flow paths through the waste. The biogas production is increased through the increased surface area available to degradation by bacteria. The disadvantage of a smaller particle size is that it allows higher packing densities, which decrease water movement, bacterial movement, and the bacterial access to substrate.

3.2.2.4 Field Capacity

Field capacity is the total amount of moisture that can be retained by a waste sample subject to gravitational pull. It is a critical measure because water in excess will form leachate and will be a major problem in landfills. Field capacity varies with the degree of applied pressure and the state of decomposition of the waste. Field capacity of residential and commercial sources is in the range of 50–60%.

3.2.2.5 Permeability

Permeability is defined as the hydraulic conductivity of compacted waste. It is an important property as it governs the movement of liquids and gases in a landfill. It depends on pore size, pore size distribution, and surface area. Permeability is inversely related to density, and hence denser refuse is less permeable. The range of permeability of MSW varies from 10^{-3} to 10^{-7} m/s.

The coefficient of permeability is represented as K

$$K = Cd^2 \frac{\gamma}{\mu} = k\frac{\gamma}{\mu} \tag{3.3}$$

where

K = coefficient of permeability
C = shape factor dimensionless
d = average size of pores
γ = Specific weight of water
μ = dynamic viscosity of water
k = intrinsic permeability

The term Cd^2 is known as the intrinsic permeability. Intrinsic Permeability values of compacted waste in a landfill are in the range 10^{-11} to 10^{-12} m^2 in the vertical direction and about 10^{-10} m^2 in the horizontal direction.

3.2.3 Chemical Properties of MSW

The chemical properties that are of importance for MSW are

1. Proximate analysis
2. Ultimate analysis
3. Fusing point of ash
4. Energy content/calorific value

These properties are important while evaluating alternative processing and recovery options and for feasibility studies of waste-to-energy processes.

3.2.3.1 Proximate Analysis of MSW

The proximate analysis of MSW is an empirical procedure for the analysis of MSW in order to list the composition in terms of the following items, all expressed as percentages by weight of MSW:

1. *Moisture:* The loss of weight when MSW is heated in an oven at 378 K for 1 hour.
2. *Volatile combustible matter:* The loss in weight when the sample is heated in a covered crucible for about 7 minutes at 1223 K (950°C) minus the weight of the moisture.
3. *Ash:* The weight of residue obtained when the sample of MSW is subjected to complete combustion in a muffle furnace at 973–1023 K.
4. *Fixed carbon:* Calculated as 100 minus the sum of the moisture, volatile matter, and ash already obtained

$$FC = 100 - (M + VCM + A) \tag{3.4}$$

where *FC*, *M*, *VCM*, and *A* are the percentages of fixed carbon, moisture, volatile combustible matter, and ash, respectively. Due to the environmental problems associated with the combustion of sulphur, it is now required to report the percentage by weight of sulphur also in the proximate analysis.

Typical values of proximate analysis are, as given in Table 3.3.

TABLE 3.3
Typical Values of Proximate Analysis of MSW

Serial Number	Components	Range (%)	Typical (%)
1	Moisture	15–40	20
2	Volatile matter	40–60	53
3	Fixed carbon	5–12	7
4	Ash	15–30	20

3.2.3.2 Ultimate Analysis of MSW

The ultimate analysis of MSW gives the composition as the percentage by weight of the various elements such as C, H, N, O, S, and the ash (Table 3.6). The ultimate analysis provides the necessary data required for material balance in the combustion of MSW. The analysis is carried out using elaborate and standard chemical techniques. The carbon dioxide, water, nitrogen oxides, and sulphur dioxide formed by burning MSW in a sufficient supply of oxygen determine, respectively, the amounts of carbon, hydrogen nitrogen, and sulphur in MSW. The CHNS Analyser is used for this as it gives rapid and precise determination of elemental composition.

The hydrogen reported in the ultimate analysis includes the hydrogen in the volatile matter and the hydrogen in the moisture present in MSW. The carbon reported in the ultimate analysis includes the fixed carbon and the carbon present in the volatile matter. The volatile matter contains complex organic substances made up of C, H, O, N, and S. The fixed carbon is the combustible residue resulting from the pyrolysis reaction at 1223 K (950°C). Oxygen and total hydrogen may be reported in the ultimate analysis, or the oxygen may be reported as equivalent 'combined water' in which case, the remaining hydrogen is the available hydrogen or net hydrogen. The available hydrogen is the only hydrogen that requires oxygen from the air for combustion. The quantity of halogens is also determined now and included in the ultimate analysis, as during combustion emission from chlorinated compounds is a matter of grave concern now. The elemental analysis is also important for the system design for biological conversions. The proper mix of waste materials for getting the suitable C/N ratio is a requirement for biological treatment.

3.2.3.3 Energy Content for Solid Waste

The energy content of the organic components is determined by using a bomb calorimeter.

Table 3.4 gives the heating values of some typical samples in kJ/kg

Using modified Dulong formula, the energy content in Btu/lb can be calculated (Tchobanoglous, 2019) which can be converted to kJ/kg:

TABLE 3.4
Heating Values of Components of Solid Waste

Components	Heating Value (kJ/kg)	
	Range	Typical
Food waste	3500–7000	4500
Paper	11,500–18,500	16,500
Cardboard	14,000–17,500	16,500
Plastics	28,000–37,000	32,500
Textiles	15,000–20,000	17,500
Municipal solid waste	9500–13,000	10,500

Source: Adapted from CS (2019).

TABLE 3.5
Typical Composition of MSW

Component	Range	Typical (% by Weight)
	Organic	
Food waste	20–65	40
Paper	10–15	12
Cardboard	2–6	5
Plastics	2–10	7
Textiles	2–6	3
Rubber	0–2	0.5
Leather	0–2	0.5
Wood	1–2	1
	Inorganic	
Glass	1–10	8
Tin cans	1–5	4
Aluminium	1–5	3
Other metals	1–4	3
Dirt, ash, etc.	1–20	13
Total		100

Source: Adapted from CPHEEO (2020).

TABLE 3.6
Sample Ultimate Analysis of the Components of the Organic Part of the MSW

Component	Weight Percent (Dry Basis)						
	Carbon	Hydrogen	Oxygen	Nitrogen	Sulphur	Ash	Total
Food waste	46.06	6.2	40.20	2.7	0.3	4.54	100
Paper	41.43	6.87	49.83	1.01	0.86	0.20	100
Cardboard	44.0	5.9	44.6	0.3	0.2	5.0	100
Plastics	60.0	7.2	22.8	–	–	10.0	100
Textiles	51.2	6.97	40.0	0.01	0.84	0.98	100
Rubber	75.72	10.76	-	1.45	0.39	11.78	100
Leather	60.5	8.1	11.0	10	0.40	10.0	100
Yard waste	48.0	6.1	40.0	3.4	0.1	3.4	100
Wood	45.69	7.57	42.24	1.89	1.01	1.6	100

$$\frac{\text{Btu}}{\text{lb}} = 145\,C + 610\left(H_2 - \frac{1}{8}O_2\right) + 40S + 10N \tag{3.5}$$

where

C = carbon, per cent by weight
H_2 = hydrogen, per cent by weight
O_2 = oxygen, per cent by weight
S = sulphur, per cent by weight
N = nitrogen, per cent by weight

Example 3.2

Compute the chemical formula of a given waste with the following analysis given in Table 3.7. The proximate analysis is as given in Table 3.8. The elemental analysis/ultimate analysis is given in Table 3.9. Calculate the energy content also.

TABLE 3.7
Analysis of Waste Sample for Example 3.2

Component	Weight (%)
Food waste	55
Paper	4.5
Cardboard	4.5
Plastics	12
Textiles	6
Rubber	0.5
Leather	0.5
Yard wastes	6
Miscellaneous	4.0
Inorganic items like glass, tin, and ash	6.5

TABLE 3.8
Proximate Analysis for Example 3.2

Component	Weight (%)
Fixed carbon	6.00
Volatile matter	55.00
Total moisture	30.00
Ash	9.00

TABLE 3.9
Elemental Analysis of the Sample for Example 3.2

Component	Unit	Quantity
Hydrogen (H)	%	4.54
Nitrogen (N)	%	2.52
Oxygen (O)	%	24.06
Phosphorus (p)	mg/kg	5206
Sulphur (S)	%	0.27
Carbon (C)	%	33.11
Cadmium (Cd)	mg/kg	0.10
Chromium (Cr)	mg/kg	17.62
Lead (Pb)	mg/kg	26.12
Arsenic (As)	mg/kg	0.10
Iron (Fe)	mg/kg	8516
Chlorine (Cl)	mg/kg	1269
Potassium (K)	mg/kg	8406
Mercury (Hg)	mg/kg	0.10

Solution

Basis: 100 kg MSW
Since the elemental analysis is given, only that data is needed. The analysis of MSW and proximate analysis data is just for information. The chemical formula with C, H, N, O, and S needs to be found out to calculate the calorific value. The data on other elements are useful when considering the different treatment options of MSW.

TABLE 3.10
Calculation of Moles of Elements in the Given Sample

Component	Unit	Quantity	Weight Kg	Atomic Weight	Mol
Hydrogen (H)	%	4.54	4.54	1.008	4.51
Nitrogen (N)	%	2.52	2.52	14.007	0.18
Oxygen (O)	%	24.06	24.06	16.0	1.50
Phosphorus (p)	mg/kg	5206	0.5206	30.974	
Sulphur (S)	%	0.27	0.27	32.065	0.0084
Carbon (C)	%	33.11	33.11	12.01	2.76
Cadmium (Cd)	mg/kg	0.10	10×10^{-6}	112.41	
Chromium (Cr)	mg/kg	17.62	17.62×10^{-4}	52.0	
Lead (Pb)	mg/kg	26.12	26.12×10^{-4}	207.2	
Arsenic (As)	mg/kg	0.10	0.10×10^{-4}	74.92	
Iron (Fe)	mg/kg	8516	0.8516	55.845	
Chlorine (Cl)	mg/kg	1269	0.1269	35.453	0.00354
Potassium (K)	mg/kg	8406	0.8406	39.098	
Mercury (Hg)	mg/kg	0.10	10×10^{-6}	200.59	

Chemical formula considering only C, H, O, S, N, and Cl

$$C_{2.76}H_{4.51}O_{1.5}S_{0.0084}N_{0.18}Cl_{0.00354}$$

Chemical formula in terms of N with moisture content

$$C_{15.3}H_{25.1}O_{8.33}S_{0.05}Cl_{0.02}N$$

Heating value can be calculated by using the modified Dulong formula (Equation 3.5).
The energy content in Btu/lb can be calculated

$$\frac{\text{Btu}}{\text{lb}} = 145\ \text{C} + 610\left(\text{H}_2 - \frac{1}{8}\text{O}_2\right) + 40\text{S} + 10\text{N}$$

where

C = carbon, per cent by weight
H_2 = hydrogen, per cent by weight
O_2 = oxygen, per cent by weight
S = sulphur, per cent by weight
N = nitrogen, per cent by weight

$$\frac{\text{Btu}}{\text{lb}} = 145 \times 33.11 + 610\left(4.54 - \frac{1}{8} \times 24.06\right) + 40 \times 0.27 + 10 \times 2.52$$

$$\text{Energy content} = 5771.78\ \text{Btu/lb} = 13425.15\ \text{kJ/kg} \qquad \text{(Answer)}$$

3.2.3.4 Plastic Material Found in MSW

Plastic material found in MSW consists of the following seven categories which can be identified by the number codes 1–7 which are moulded into the bottom for easy identification (Figure 3.1).

FIGURE 3.1 Plastic material identification number code

1. Polyethylene terephthalate (PETE)
   Soft drink bottles, mineral water bottles, fruit juice containers, and cooking oil containers
2. High-density polyethylene (HDPE)
   Milk jugs, containers for cleaning agents, laundry detergents, bleaching agents, shampoo bottles, washing, and shower soaps.
3. Polyvinyl chloride (PVC)
   Trays for sweets, fruits, plastic packings like bubble foil, and food foil to wrap the foodstuff.

4. Low density polyethylene (LDPE)
   Crushed bottles, shopping bags, highly resistant sacks, and most of the wrappings.
5. Polypropylene (PP)
   Furniture, toys, luggage, bumpers, lining, and external borders of cars
6. Polystyrene (PS)
   Toys, hard packings, refrigerator trays, cosmetic bags, costume jewellery, CD cases, and vending cups
7. Other plastics
   Includes all other plastics like polycarbonates, acrylic, polyacetals, fibres, nylon, fiberglass, etc.

3.2.3.5 Fusing Point of Ash

The fusing point of ash is defined as the temperature at which the ash from the burning of MSW will form a clinker by fusion and agglomeration. Typical values are in the range 1373–1473 K for MSW.

3.2.4 Biological Characteristics of MSW

The organic fraction of almost all components of MSW can be converted biologically to gases and relatively stable organic and inorganic solids and liquids. The production of odours and the generation of flies are also related to the putrescible nature of the organic materials found in MSW. The organic fraction of most MSW can be classified as below (excluding plastic, leather, and rubber compounds).

1. *Carbohydrates*
   Carbohydrates are the main constituent of food and yard wastes. They include sugars and polymers of sugars such as starch and cellulose with the general formula $(CH_2O)_n$. They are easily biodegradable and methane, carbon dioxide, and water are the products formed. Degradation of carbohydrates attract flies and rats and care should be taken to eliminate it.
2. *Natural fibres*
   Natural compounds like cellulose and lignin are the main constituents in this category. They are resistant to biodegradation. They are found in paper and paper products and also in food waste and yard waste. They are highly combustible and are suitable for incineration. The cellulose content of paper, cotton, and wood products are 100%, 95%, and 40%, respectively. They have very high calorific values also.
3. *Lipids* include fats, oils, and waxes, which are esters of alcohols and long-chain fatty acids. Though they are biodegradable, the rate of biodegradation is slow.
4. *Proteins*, which are composed of chains of amino acids, constitute 5–10% of the solid waste. These compounds contain carbon, hydrogen, oxygen, and nitrogen. Proteins decompose to amino acids. Partial decomposition produces amines which have a very unpleasant smell.

3.2.4.1 Breeding of Flies

Fly breeding is an important consideration for onsite storage of waste and in waste handling plants. In warm climates, this is a serious issue. The fly breeding cycle is as follows (Table 3.11).

TABLE 3.11
Fly Breeding Cycle

Stage of Growth	Time Required
Eggs develop	3–12 h
First stage of Larval period	20 h
Second stage of larval period	24 h
Third stage of larval period	3 days
Pupal stage	4–5 days
Total	9–11 days

Removing maggots from containers is difficult, even when emptied, as the maggots in the emptied cans will develop into flies. Maggots also crawl into the surroundings and develop into flies which is a very serious problem for the neighbourhood of waste treatment plants.

3.2.4.2 Biodegradability of Organic Wastes

There are many biological processes like composting, anaerobic digestion, and mechanical biological treatment processes for stabilising the organic part of MSW. Stability is defined as the extent to which biodegradable waste has decomposed. There are different methods suggested for measuring this, as a proper knowledge of the biodegradability is key to the appropriate design of the treatment facilities.

One measure suggested is the volatile solids (VS) content determined by ignition at 823 K (550°C). This can be misleading, as some organic material may be highly volatile but low in biodegradability like some plant trimmings or newsprint. The lignin content is another suitable method that is suggested. It is defined as

$$BF = 0.83 - 0.028\ LC \tag{3.6}$$

where

BF = biodegradable fraction expressed on a volatile solids (VS) basis
LC = lignin content of the VS expressed as a percent of the dry weight
0.83, 0.028 are empirical constants (Tchobanoglous, 2019)

Respiration indices are one of the most suitable methods for determining the stability of organic wastes. Dynamic respiration indices (DRI) expressed as average oxygen uptake rate for one hour and 24 hours indicated as DRI_{1h} and DRI_{24h} are found useful as biodegradability measure.

TABLE 3.12
BF Based on Lignin Content Computed Using Equation 3.6

Component	Volatile Matter % VS	Lignin Content LC	Biodegradable Fraction
Newsprint	98	22	0.214
Cardboard	94	13	0.466
Food waste	96	0.5	0.816
Yard waste	88	4	0.718

3.2.4.3 Odour

MSW doesn't have any odour as such. Odour starts developing when the organic part of the MSW starts putrefying. Odour from MSW is due to complex mixtures of a large number of volatile organic compounds. Odour concentration is the measure of the detectability of odour. Odour intensity is defined as the perceived magnitude of a stimulus. Odour intensity and offensiveness are subjective measures of the strength and unpleasantness of an odour as perceived by a panel of people. Total odour can be described by both quantitative and qualitative means based on strength, intensity, character, and hedonic tone. The relative strength of an odour can be determined based on a relationship between the volumes of fresh air required to dilute a volume of odorous air to a level where odour is not detectable. This relationship is termed dilution threshold.

3.2.5 Properties of Hazardous Waste in MSW

Hazardous waste is waste that must be handled properly to avoid damage to human health or environment. Waste can be hazardous if it is toxic, reacts violently with other chemicals or if it is corrosive.

3.2.5.1 Hazardous Wastes in MSW

MSW will contain some hazardous wastes from households and shopping complexes. These hazardous wastes will go into landfills or incinerators unless segregated separately. Management of hazardous wastes is very important as it will cause damage to soil, air, and water if it comes into contact. Even trace amounts of hazardous wastes if present in compost will make the entire compost unusable as it will damage the soil and if absorbed by plants and crops it will be dangerous and the problem of bioaccumulation also is a reality. If MSW that is incinerated contains any type of hazardous materials, then the flue gas will have these components and will pollute the air. The ash resulting from this which goes to landfill will have heavy metal toxicity. The presence of hazardous waste will have multiple issues like getting into landfill leachate, landfill gas, and it can be the hazardous waste itself or as a result of chemical or biological reactions of the hazardous waste inside the landfill. A waste in MSW is considered hazardous based on whether it is reactive, ignitable, toxic, corrosive, or carcinogenic. From a health-related aspect, more properties are considered like infection-prone, irritant, mutagenicity, radioactivity, and teratogenicity.

3.2.5.2 Reactivity

A solid waste is said to be reactive if a sample of waste has the following characteristics:

a. It is normally unstable and readily undergoes violent changes without detonating.
b. It reacts violently with water.
c. It forms potentially explosive mixtures with water.
d. When mixed with water, it generates toxic gases, vapours, or fumes in a quantity sufficient to present a danger to human health or the environment.
e. It is a cyanide or sulphide-bearing waste, when exposed to a pH between 2 and 12.5, can generate toxic gases, vapours, or fumes in a quantity sufficient to present a danger to human health or the environment.
f. It is capable of detonation or explosive decomposition or reaction at standard temperature and pressure.
g. It is capable of detonation or explosive reaction if it is subjected to a strong initiating source or if heated under confinement.

3.2.5.3 Flammable (Ignitability)

A solid waste sample is said to be ignitable if it has any of the following properties.

a. It is a liquid other than an aqueous solution containing less than 24% alcohol by volume and has a flash point less than 333 K.
b. It is not a liquid and is capable, under standard temperature and pressure, of causing fire through friction, absorption of moisture, or spontaneous chemical changes, and when ignited, burns so vigorously, and persistently that it creates a hazard.
c. It is an ignitable, compressed gas.
d. It is an oxidiser.

3.2.5.4 Corrosivity

A solid waste is said to be corrosive if a waste sample has either of the following properties.

a. It is aqueous and has a pH ≤ 2 or ≥ 12.5.
b. It is a liquid that corrodes steel at a rate of >6.35 mm per year at a test temperature of 328 K.

3.2.5.5 Toxicity

The regulatory levels of the toxic materials are listed in Table 3.15. These properties are primarily based on the safety and health-related effects of these wastes to humans, animals, and plants in general. Hazardous wastes are classified based on whether they are persistent or nonpersistent. An example of persistent organic waste is pesticide dichloro diphenyl trichloro ethane (DDT). DDT is harmful to all biota as it is non-biodegradable. It is a notorious insecticide that was considered to be a great scientific breakthrough and widely used from 1940 to 1970 for use in household, agriculture, and public health vector control projects. It tends to become more concentrated as it moves up the food chain. Now it is listed as a probable carcinogen. It was widely used in World War II to protect soldiers from insect-borne diseases like malaria and typhus. In 1962, Rachel Carson in the book *Silent Spring* alerted the authorities concerned and the public at large that DDT bioaccumulates and biomagnifies up the food chain and raised concerns of the lasting effects on wildlife and on humans. It was noticed that a high level of DDE which is a metabolite of DDT, in certain birds caused its egg shells to thin and live offspring could not be produced. The bald eagle was one species affected. But after banning the use of DDT resulted in the species recovery hitherto unknown in history. Stockholm Convention on Persistent Organic Pollutants (POP) 2008 was adopted by 160 countries and committed to eliminate 12 POPs of greatest concern including DDT. The 12 POPs are named as the dirty dozen. Though DDT was banned in the United States in 1972 and many countries followed suit, it is still permitted to be used in some Asian and African Countries for vector control. POPs can travel thousands of miles when they evaporate to air by attaching to airborne particles and have been detected even in pristine Arctic region as detected by satellite images. The POPs are listed in Table 3.13.

TABLE 3.13
List of POPs – The Dirty Dozen

POP	Use/Source
Aldrin	Insecticide for crops such as corn and cotton
Dieldrin	Used for termite control
Chlordane	Insecticide on crops, vegetables, small grains, potatoes, sugarcane, fruits, nuts and cotton Termite control, garden pest control and home lawn pest control
DDT	Insecticide used in agricultural crops, Insect control for prevention of malaria and typhus
Endrin	Insecticide used on crops like cotton and grains. Used to control rodents
Mirex	Insecticide used to control termites, fireants, mealybugs. It is a fire retardant used in rubber, plastics and electrical products
Heptachlor	Insecticide used in soil to control termites and soil insects. Used to combat malaria
Hexachloro benzene	Fungicide used for seed treatment. An industrial chemical with applications in fireworks, ammunitions, synthetic rubber, etc. Unintentionally produced during combustion and manufacture of certain chemicals. It is also an impurity present in certain pesticides
Polychlorinated biphenyls (PCBs)	Used in many industrial processes and purposes, including in electrical transformers and capacitors, as heat exchange fluids, paint additives, in plastics. Unintentionally produced during combustion
Toxaphene	Insecticide used to control pests on crop and livestock. To kill unwanted fish in lakes
Dioxins	Unintentionally produced during most forms of combustion including burning of municipal and medical wastes, backyard burning of thrash, and industrial processes.
Polychlorinated dibenzofurans	Found as contaminants in certain herbicides, wood preservatives and PCB mixtures

3.2.5.6 Dioxins and Polychlorinated Biphenyls (PCBs)

Dioxins are not manufactured for any particular use, rather they are produced as a contaminant byproduct while manufacturing or burning chlorophenols. No dioxin is formed naturally in the environment. Chlorodioxin has nearly 20 isomers identified. The most poisonous of all synthetic chemicals is 2,3,7,8-tetrachlorodibenzo-p-dioxin

TABLE 3.14
Hazardous Wastes from Households

Product	Concern	Disposal
	Household Cleaners	
Abrasive scouring powders	Corrosive	Hazardous waste facility
Aerosols	Flammable	Hazardous waste facility
Ammonia and ammonia-based cleaners	Corrosive	Hazardous waste facility or dilute small amounts
Chlorine bleach	Corrosive	Hazardous waste facility or dilute small amounts
Drain openers	Corrosive	Hazardous waste facility
Furniture polish	Flammable	Hazardous waste facility
Glass cleaners	Irritant	Dilute small amounts
Outdated medicines	Hazardous to others in the family	Dilute small amounts and flush down toilets
Oven cleaner	Corrosive	Hazardous waste facility
Shoe polish	Flammable	Hazardous waste facility
Silver polish	Flammable	Hazardous waste facility
Spot remover	Flammable	Hazardous waste facility
Toilet bowl cleaner	Corrosive	Hazardous waste facility
Upholstery and carpet cleaner	Flammable/corrosive	Hazardous waste facility
	Personal Care Products	
Medicated shampoos	Poison	Dilute small amounts and flush down toilets
Nail polish remover	Poison, flammable	Hazardous waste facility
Rubbing alcohol	Poison	Dilute small amounts and flush down toilets
	Automotive Products	
Car batteries	Corrosive	Recycling Centre
Diesel, gasoline, kerosene, waste oil	Flammable	Recycling Centre
	Paint Products	
Paints, enamels	Flammable	Hazardous waste facility
Paint Solvents, Thinners	Flammable	Hazardous waste facility
	Miscellaneous Products	
Batteries	Corrosive	Hazardous waste facility
Pool acids and chlorine	Corrosive	Hazardous waste facility
Pesticides, herbicides, Fertilisers, insecticides	Poison	Hazardous waste facility

Source: Adapted from Davis (2022).

(TCDD). TCDD contamination is reported in particulate matter from commercial and domestic combustion processes. TCDD is a stable compound and is a white crystalline solid which has very limited solubility in water, 0.2–0.6 ppb. It is thermally degraded at over 973 K (700°C) and undergoes photochemical degradation under UV light in the presence of a hydrogen-donating solvent like a solution of olive oil in cyclohexanone. The adverse health effects in people with dioxin exposure came to light when birth defects in newborns were found in South Vietnam where Agent Orange was used in warfare. Agent Orange is an herbicide and defoliant and US military sprayed it from 1962 to 1971 during Vietnam War to remove the leaves of trees and other dense tropical foliage that provided enemy cover. Nearly three lakh veterans died of exposure to agent orange. Even now it is poisoning Vietnamese people. Studies of dioxin contamination in soil and sediment samples in Vietnam show that a major route of current and past exposures is from the movement of dioxin from soil into river sediments, then into fish, and from fish consumption into people. Much of the damage can never be repaired is the conclusion from various studies. Animal toxicological investigations revealed that TCDD is a carcinogen, a teratogen, a mutagen, an embryo-toxin, causes severe skin disorders, and affect immune responses in animals. It is a persistent organic material and it bioaccumulates in humans and aquatic organisms. Dioxins even in trace amounts are carcinogens and may cause disruption of regulatory hormones, reproductive and immune system disorders, and abnormal foetal development.

PCBs are produced by chlorination of a biphenyl molecule. It is composed of ten possible forms and has more than 200 isomers. Monochlorobiphenyl, dichlorobiphenyl, and trichlorobiphenyl are examples. The isomers are 3-chlorobiphenyl, 2,4'-dichlorobiphenyl, 2,4,4',6-tetrachlorobiphenyl, etc., depending on the substitution sites of the biphenyl portion of the molecule. PCBs were extensively used in the period from 1930 to 1970, when the harmful effects were not known. It was used in paints, inks, dusting agents, and pesticides. Toxic effects were first noticed in exposed workers. The chlorine content of the PCBs varied from 18 to 79%. Mixture of PCBs was used as coolant/dielectric for transformers and capacitors, as a heat transfer fluid and as protective coatings for wood where low flammability was desirable.

TABLE 3.15
Hazardous Waste Compounds from Commercial, Industrial and Agricultural Activities That Are Found in MSW

Name	Symbol/ Formula	Regulatory Level (mg/L) For Toxicity	Concern	Long-Term Effect
		Metals		
Barium	Ba	100.0	Flammable at room temperature when in powder form	Increased blood pressure and nerve block

Continued

TABLE 3.15
Continued

Name	Symbol/ Formula	Regulatory Level (mg/L) For Toxicity	Concern	Long-Term Effect
Cadmium	Cd	1.0	Flammable in powder form. Toxic if inhaled in powder form or fumes Carcinogen Soluble compounds of cadmium are highly toxic.	Concentrates in the liver, kidneys, pancreas and thyroid. Hypertension suspected effect
Chromium	Cr	5.0	Hexavalent chromium compounds are carcinogenic and corrosive on tissue.	Skin sensitisation and kidney damage
Lead	Pb	5.0	Toxic if inhaled in powder form or fumes	Brain, nervous system and kidney damage. Birth defects
Mercury	Hg	0.2	Highly toxic by inhalation of fume or vapour or absorption by skin.	Toxic to central nervous system. May cause birth defects
Silver	Ag	5.0	Toxic metal	Permanent grey discolouration of skin, eyes and mucous membrane
		Nonmetals		
Arsenic	As	5.0	Carcinogen and mutagen	Causes fatigue and loss of energy. Dermatitis
Selenium	Se	1.0		Red staining of fingers, teeth and hair. General weakness, depression, irritation of nose and mouth
		Organic Compounds		
Benzene	C_6H_6	0.5	Carcinogen. Highly toxic, flammable, dangerous fire risk	
Toluene	$C_6H_5CH_3$	200.0	Toxic by ingestion, inhalation and skin absorption Irritant to skin and eyes. flammable, dangerous fire risk	

Continued

TABLE 3.15
Continued

Name	Symbol/ Formula	Regulatory Level (mg/L) For Toxicity	Concern	Long-Term Effect
		Halogenated Compounds		
Chlorobenzene	C_6H_5Cl	100	Toxic by inhalation and skin contact. Fire risk	
Vinyl chloride	CH_2CHCl	0.2	Extremely toxic and hazardous by all types of exposure. Carcinogen	
		Pesticides, Herbicides, Insecticides		
Endrin	$C_{12}H_8OCl_6$	0.02	Toxic by skin absorption and inhalation, Carcinogen	
Lindane	$C_6H_6Cl_6$	0.4	Toxic by skin absorption, ingestion and inhalation, Carcinogen	
Methoxychlor	$C_{16}H_{15}Cl_3O_2$	10.0	Toxic	
Toxaphene	$C_{10}H_{10}Cl_8$	0.5	Toxic by ingestion, skin absorption and inhalation, Carcinogen	
Silvex	$C_9H_7Cl_3O_3$	1.0	Toxic material Use restricted now	

Source: Adapted from Davis (2022), MoEF (2016).

3.2.5.7 Nonpersistent Organic Wastes

Nonpersistent pollutants are biodegradable and can be readily broken down by natural processes. Toxic problems associated with nonpersistent pollutants occur at the point of release and at the exposure time itself. Examples are biodegradable pesticides like organophosphates, triazines, detergents, and waste oils.

3.2.5.8 Half-Life Period

The environmental persistence of a hazardous chemical is usually expressed by its half-life period. Half-life is the time required for half the reaction to be completed. The time required for half the reaction to be over is independent of the initial concentration of the reactant. For a first-order reaction, whatever be the initial concentration, it will be reduced to half its value independent of the initial concentration in its half-life period. The hazardous waste materials that are present in MSW are in very low concentrations and the decomposition reactions can be adequately represented by first-order kinetics.

The kinetic expression for first-order reaction is represented as

$$-\frac{dC_A}{dt} = kC_A \tag{3.7}$$

C_A = concentration at time t, t = time k = first-order reaction rate constant
Equation 3.7 can be integrated with the following limits

$$t = 0 \quad C_A = C_{A0} \qquad t = t \quad C_A = C_A$$

$$\int_{C_{A0}}^{C_A} -\frac{dC_A}{C_A} = \int_0^t kt$$

$$\text{Which gives} \quad -\ln\frac{C_A}{C_{A0}} = kt \tag{3.8}$$

$$\text{For half-life } t = t_{1/2} \quad \frac{C_A}{C_{A0}} = 0.5$$

Substituting in Equation 3.8, we get

$$t_{1/2} = -\frac{\ln 0.5}{k} = \frac{0.6932}{k} \tag{3.9}$$

Equation 3.9 can be used to determine the time for concentration reduction to any value or alternatively to find the concentration at any time like $t_{1/4}$, $t_{1/8}$, $t_{3/4}$, and $t_{7/8}$ if the value of rate constant k is known.

Equation 3.9 can also be expressed in terms of fractional conversion of X_A of a given reactant A

$$C_A = C_{A0}(1 - X_A) \tag{3.10}$$

Equation 3.8 becomes

$$-\ln(1 - X_A) = kt \tag{3.11}$$

If the reaction is second order, the rate expression is

$$-\frac{dC_A}{dt} = k_2 C_A^2 \tag{3.12}$$

Where k_2 is the second-order rate constant
In general, for an nth-order reaction

$$-\frac{dC_A}{dt} = k' C_A^n \tag{3.13}$$

Integrating for $n \neq 1$ gives

$$C_A^{1-n} - C_{A0}^{1-n} = k'(n-1)t \tag{3.14}$$

And half-life of the reaction $t_{1/2}$ is obtained as

$$t_{1/2} = \frac{2^{n-1} - 1}{k'(n-1)} C_{A0}^{1-n} \tag{3.15}$$

Example 3.3

The half-life of a hazardous substance A is 24 hours. Determine $t_{1/3}$

Solution

Half-life is given by

$$t_{\frac{1}{2}} = \frac{0.6932}{k}$$

$$\text{Given} \quad t_{1/2} = 24 \text{ h} \quad \text{So,} \quad k = \frac{0.6932}{24} = 0.0289 \text{ h}^{-1}$$

From Equation 3.11

$$-\ln(1 - X_A) = kt$$

where X_A is the fractional conversion

$$X_A = 0.333$$

$$-\ln 0.667 = kt_{1/3}$$

$$t_{1/3} = -\frac{\ln 0.667}{0.0289} = 14.01 \text{ h}$$

Time required for one-third depletion = $t_{1/3}$ =14.01 h (Answer)

Example 3.4

Phenol is spilled in a waste water treatment pond. Find the time required for the concentration to be reduced to half if the reaction rate constant for phenol removal is 6.65×10^{-3} h^{-1}

Solution

The time required for concentration reduction to half = $t_{1/2}$

$$t_{1/2} = \frac{0.6932}{k} = \frac{0.6932}{6.65 \times 10^{-3}} = 104 \text{ h}$$ **(Answer)**

Example 3.5

0.2 g of radium (^{226}Ra) emits 0.72×10^{10} α particles per second. Calculate its half-life.

Solution

$$\text{Rate} = -\frac{dC_A}{dt} = kC_A = 0.72 \times 10^{10} \text{ disintegrations per second}$$

$$k = \frac{0.72 \times 10^{10}}{C_A}$$

Replace C_A by N in this case where N = number of atoms of radium

$$N = \frac{0.2 \times 6.023 \times 10^{23}}{226} = 5.33 \times 10^{20}$$

$$k = \frac{0.72 \times 10^{10}}{5.33 \times 10^{20}} = 0.1351 \times 10^{-10} \text{ s}^{-1}$$

Half-life $t_{1/2}$ is given by Equation 3.9

$$t_{1/2} = -\frac{\ln 0.5}{k} = \frac{0.6932}{k}$$

Substituting the value of k

$$t_{1/2} = -\frac{\ln 0.5}{k} = \frac{0.6932}{k} = \frac{0.6932}{0.1351 \times 10^{-10} \times 3600 \times 24 \times 365} = 1627 \text{ years}$$

(Answer)

The half-lives of persistent organic compounds because of their slow rate of decomposition can vary anywhere from 100 years to 7000 years (e.g. tetrachloromethane, 7000 years; trichloroethane, 170 years). Combustion process is an effective method of disposing of hazardous material in MSW. The combustion has to be complete or else toxic compounds will be produced. The long-term management of these compounds in landfills is difficult due to their long half-lives. Chemical transformations with metals also take place and highly toxic materials like methylmercury, dimethyl arsine (C_2H_7As) is produced in landfills. Biological transformation of mercury will continue for years. Gaseous products will be formed from hazardous wastes by vapourisation and as gaseous products formed by reaction or from hazardous wastes bound to soil volatilises. Segregation at the point of generation and sending it to approved disposal facilities is the best method to avoid contamination in soil, water, and air.

3.3 SEGREGATION, COLLECTION, AND TRANSPORTATION OF MSW

The segregation, collection, and transportation strategy are dependent on the means of processing. The factors influencing the strategy include

- Quantity of solid wastes to be processed
- Process options
- Recyclability

- Potential for value-added products
- Feasibility of processing at a large scale by using solid wastes from multiple sources

Irrespective of the variable factors, the separation of hazardous wastes and biomedical wastes is mandatory in all cases. The segregation normally is done both at the source and at the processing centre. The segregation at source itself depends on the method adopted for waste processing. There are different approaches to segregation at source in solid waste management, namely single bin, two bin, three bin and four bin system.

3.3.1 Single Bin System

In single bin system, all the wastes are collected in the same bin and transferred to the processing centre. However, even in single bin system, there could be restrictions on what can be dumped into the bin and what should not be. The wastes collected in the bin is sorted at the processing centre as per the processing needs.

3.3.2 Two Bin System

There are different types of two-bin systems. In one case of two bin system, the wastes are classified into just recyclables and non-recyclables. In this case, householders are required to keep the recyclables separate and leave the rest of the waste in a single bin. The recyclables are sorted at the processing centre and sent to appropriate agencies for processing. This type of sorting is required since recyclable wastes include plastics, glass, metals, paper, and cardboard. Each item has to be stored separately before disposal or sale. The mixed wastes would be sorted for degradable and non-degradable wastes if composting is an option. Further segregation of both degradable wastes and non-degradable wastes would be required if the economics of scale justify such sorting for manufacturing value-added products. The simplest of examples include manufacture of door mats with waste clothes or dishwashing powder with lemon peels. If such mixed wastes are collected for processing in a waste to energy plant (WTE) or for simply disposed in sanitary landfill, further segregation is not required. If there are several small towns storing the mixed waste to be transported to a single point for either as raw material for a WTE plant or sanitary landfill, there should be measures for odour and vector control in the storage area. A bin with the same colour lid as the body is meant for mixed waste and one with light-coloured lid is meant for recyclable wastes. Householders have to strictly adhere to the system of segregation.

The approach in this case is to collect the biodegradable wastes separately and process that to manufacture compost. The non-degradable wastes would be sorted out at the processing centre to separate recyclable components, saleable components, components which may have use as fuel in some factories and which may go to a sanitary landfill. In this case, the householder has to separate the biodegradable wastes and non-biodegradable wastes. Alternatively, coloured bins like red coloured bins are used for nonbiodegradable wastes and green coloured bins are used for collecting biodegradable wastes.

3.3.3 Three-Bin Method

Segregation as recyclable, biodegradable, and non-biodegradable at source itself is done here. In this method there would be three bins. This is a practice generally adopted at public places like bus stations, transport terminals like airports, and train stations.

3.3.4 Four-Bin Method

This method of segregation is also used nowadays. Blue bins for dry recyclable materials, green bin for wet biodegradable wastes, red bin for hazardous and medical waste and yellow bin for dry waste which cannot be recycled or composted.

3.3.5 Waste Collection and Transportation Models for Households

The local self-government (panchayat, municipality, corporation, or similar government department) can recruit a squad and entrust them with the waste collection. The squad is given training by experts. Each squad consists of four to ten people and they are allotted up to 250 houses depending on the population of the town. Each household is to give a monthly user fee. The households have to segregate and keep the waste for handing over. A successful model that is adapted is shown in Table 3.16. Biodegradable waste is to be collected on a daily basis and sent to the processing plant. Plastic waste is collected on a monthly basis. By this schedule, the segregation becomes automatic. The squad can sell the items that have a market and the rest can be handed over to the facility for dealing with such waste management by the government or such authorities. By this process, the waste collectors are getting an income and there will not be any ragpickers. The squad is to be provided with appropriate vehicles for transportation of the waste and also facilities for waste storage till such time that they can handover the waste to the concerned authorities. A few examples of the vehicles are shown in the Figures 3.2 and 3.3.

The transportation is generally carried to the processing centre in dedicated vehicles. The design of collection cum transportation vehicles would be based on the segregation practice at source and processing system at the receiving centre. Basically,

FIGURE 3.2 Vehicle for transportation of MSW. Photo taken by the author

FIGURE 3.3 Vehicle with compartments for transportation of segregated MSW. Photo taken by the author

TABLE 3.16
Schedule for Non-biodegradable Waste Collection

Month	Type of Waste
January, September	Plastic waste, E waste, paper and cardboard
February, October	Plastic waste, cloth waste
March, November	Plastic waste, Toxic E waste
April, December	Plastic waste, shoes, bags, upholstery, thermocol, carpet
May	Plastic waste, glass and porcelain
June	Plastic waste, Tyres
July	Plastic waste, scrap metal
August	Plastic waste, Medicine strip

there are two variants. One is to have different compartments within the vehicle and the other is to load the vehicle with bins carrying segregated wastes. The use of electric vehicles charged with solar power is the best option as it is more environmentally friendly. Some local governments have completely shifted to the use of such vehicles.

The schedule given in Table 3.16 is successfully implemented in some local bodies. In public spaces like bus stands, community halls, places of worship, shops, establishments, and hotels separate bins are to be provided for each type of waste. Hospital waste management is to be done as per the existing rules and regulations for biomedical waste.

3.3.6 Collection of Waste from Shops and Establishments

Doorstep collection of the waste is to be done as and when the shop is open. User fee should be fixed based on the weight of the waste. The waste storage has to be in separate bins as per norms. Usually, the recyclable waste from shops will be dry and will fetch an income for the squad members. Large commercial complexes, hotels and restaurants, markets, community halls and marriage halls, garden waste, institutional sources like schools, offices, commercial sources like service stations, warehouses, farms like poultry, municipal sources like street sweepings, leaves, sewage treatment solids, and water treatment solids should make arrangements with the local self-government for the management of the waste generated.

TABLE 3.17
Data of a Typical Small Town

Current population	13,000
Floating population per day	3500
Number of houses	3500
Number of commercial establishments	541
Number of wards	15
Number of streets	75
Area	12 sq.km
Source Segregation	100%
Waste Diversion	100%
Organic waste collection per day	1550 kg
Food waste collection per day	50 kg
Meat waste collection per day	100 kg
Recycle waste collection per day	980 kg
Landfill waste collection per day	320 kg
Total waste collection per day	3.0 MT

Vehicles used for waste movement corresponding to Table 3.17:

Tractor with tipper:	1 Number
Battery Operated Vehicle:	4 Numbers
Push Cart:	18 Numbers

3.4 GENERAL PROCESSING METHODS – COMPOSTING

Composting was used by farming community ever since people changed from hunter gatherers to settlers after getting to know of agriculture. So, the history of composting can be said to be very long and rich and the practice of composting is still influencing the current thinking of using organic farming. Ancient Indian and Chinese writings have references to soil enriching with the use of manures and organic material. Evidences of the use of organic material for improving soil condition by ancient civilisations were discovered accidentally as well as by archaeological excavations. Though many civilisations rose and fell, knowledge of composting, and methods to improve soil fertility were passed on to future generations. As writing became common such experiences were recorded. References to the use of animal dung mixed with plant waste for agriculture is mentioned in the book De Agri Cultura (Concerning the Culture of the Fields) written by Roman General Cato the elder. Early Egyptians composted with vermi. Cleopatra even went to the extent of declaring earthworms as sacred recognising its contribution to agriculture. The early Arabic writer Ibn al Awan describes the use of compost in his book Kitab al Falahah (Book on Agriculture). In 1905, Sir Albert Howard worked as an agronomist in India conducting research on farming practices and composting in the Indian subcontinent. The knowledge he acquired is the foundation for present-day understanding of composting. Soil health is linked to plant health and animal health. Compost has many uses beyond soil fertility. It can be used to remediate soils contaminated with hydrocarbons, heavy metals, or chemicals. It can also be used to filter contaminants from stormwater runoff.

3.4.1 Impacts of Untreated Solid Wastes

The organic content of municipal solid waste (MSW) leads to pollution of the environment. It decomposes, generating foul-smelling and easily combustible gases and causing fire hazards. To ensure a safe disposal of the MSW it is essential to minimise environmental pollution and fire hazards. There are several processing methods available for this purpose. Odour develops in onsite storage like collection centres, transfer stations, and dumping grounds and in warm climate it is even more significant. The foul smell can be attributed to the production of hydrogen sulphide under anaerobic conditions from decomposable organic compounds. The breeding of flies is another important problem for organic waste storage in warm climates. Flies develop in a period of 9–11 days and it is difficult to remove. It will spread to the surrounding environment and the maggots will be crawling around.

3.4.2 Composting Process

Composting process is one of the most commonly used methods of reusing the organic decomposable part of MSW and this results in the production of a stable product – compost. Depending upon its quality, compost can be used as a low-grade manure and soil conditioner. The process of composting results in conservation of natural resources and is an important processing method, especially in agricultural and horticultural sectors. Decomposition and stabilisation of organic waste matter is a natural phenomenon. Composting is an organised method of producing compost manure by adopting this natural phenomenon. Compost is particularly useful as an organic manure which contains plant nutrients (nitrogen, phosphorus and potassium) as well as micro nutrients which can be utilised for the growth of plants. If required, usage in conjunction with chemical fertilisers is an option for optimising the results. The purpose of composting is the following:

a. To transform the biodegradable part of MSW to a biologically stable material and thus reduce the bulk volume of the material. The quantity of compost is less than 25-30% of the feed material.

b. To destroy the pathogens, insect eggs, and other unwanted organisms present in the MSW.
c. To destroy the weed seeds.
d. To retain the nutrients nitrogen, phosphorus, and potassium present.
e. To produce a good product that supports plant growth.

3.4.3 Separation of Non-Degradable and Recyclable Solid Wastes

The waste is fed to a slowly moving (5 m/min) conveyor belt and the non-decomposable material such as plastics, glass, metals are manually removed by workers standing on either side of the conveyor belt. The workers are provided with hand gloves to manually remove the material from the moving belt (the thickness over the belt is kept less than 0.15 m) and the removed material is stored separately.

3.4.4 Mechanical Size Reduction

The waste is then subjected to size reduction when the surface area per unit weight is increased for faster biological decomposition. Size reduction also helps in reducing fly breeding in the decomposing mass. This is commonly carried out either in hammer mills or Rasp mills. Hammer mills are high speed (600–1200 revolutions per minute) compact machines but consume large energy. Rasp mills are slow-moving large units that require lesser energy. The capital cost of a hammer mill is less but its operating cost is more than that of a Rasp mill, mainly due to the larger energy requirement as well as more frequent replacement or re-tipping of hammers.

3.4.5 Types of Composting – Aerobic and Anaerobic

During aerobic composting aerobic microorganisms oxidise organic compounds to carbon dioxide, nitrite, and nitrate. Carbon from organic compounds is used as a source of energy while nitrogen is recycled. Due to exothermic reaction, temperature of the mass rises.

3.4.5.1 Aerobic Composting

$$\text{Organic matter} + O_2 + \text{Nutrients} + \text{Microorganisms} \rightarrow \text{New Cells} + \text{Organic matter} + CO_2 + H_2O + NH_3 + SO_4^{2-} + \text{Heat} \quad (3.16)$$

The components of the organic fraction in MSW are proteins, amino acids, lipids, carbohydrates, cellulose, lignin, and ash. The components of the organic matter in the product are mainly cellulose, lignin, and ash. When plants decompose, NH_3 is produced. Ammonia is converted to nitrate ions NO_3^+ by reacting with oxygen in a two-step process.

$$2NH_3 + 3O_2 \rightarrow 2NO_2 + 2H^+ + 2H_2O$$

$$2NO_2 + O_2 \rightarrow 2NO_3$$

Equation 3.16 can be modified as

$$\text{Organic matter} + O_2 + \text{Nutrients} + \text{Microorganisms} \rightarrow \text{New Cells} + \text{Compost} + CO_2 + H_2O + NO_3^- + SO_4^{2-} + \text{Heat} \quad (3.17)$$

3.4.5.2 Anaerobic Composting

During anaerobic process, the anaerobic microorganisms, while metabolising the nutrients, breakdown the organic compounds through a process of reduction. A very small amount of energy is released during the process and the temperature of composting mass does not rise much. The gases evolved are mainly methane and carbon dioxide. An anaerobic process is a reduction process and the final product is subjected to some minor oxidation when applied to land. Composting of organic matter by anaerobic process can be represented by the following equation.

$$\text{Organic matter} + H_2O + \text{nutrients} \rightarrow \text{New Cells} + \text{Organic matter} + CO_2 + CH_4 + NH_3 + H_2S + \text{Heat} \quad (3.18)$$

The new cells formed will become dead cells and it will be a component of the compost for both aerobic and anaerobic process. Aerobic composting stabilises in a short time as compared to anaerobic composting which requires longer time for stabilisation. Anaerobic composting needs larger land space and the gases generated pose odour problems. Aerobic composting needs lesser land space and no odorous gases are produced. So aerobic composting is more preferable.

Aerobic microorganisms decompose organic material to obtain energy and material they need for growth and reproduction. In composting, there is an active composting period and a curing period. During the active composting period there is vigorous microbial activity and the readily degradable biomass degrades. During curing period, there is low level microbial activity and further decomposition and stabilisation is achieved. Curing period lasts one to six months. The organic acids and decay resistant compounds decompose and humic compounds and nitrate nitrogen are formed. The compost temperature comes to the ambient level.

3.4.6 Windrow Composting

The organic material present in municipal waste can be converted into a stable mass by aerobic decomposition. The aerobic bacteria as well as other microorganisms in the degradable solid wastes facilitate the reaction. Aerobic microorganisms oxidise organic compounds to carbon dioxide and oxides of nitrogen and use carbon from organic compounds as a source of energy, while nitrogen is recycled. Due to exothermic reactions, temperature of mass rises. Windrows method is preferred in areas/regions where higher ambient temperatures are available. Typically, each windrow will be of dimensions 3 m long × 2 m wide × 1.5 m high, with a total volume not exceeding 9.0 m^3. Each windrow would be turned on 6th & 11th days outside to the centre to destroy insect's larvae and to provide aeration. On 16th day, windrow would be broken down and passed through manually operated rotary screens of about 25 mm square mesh to remove the oversize material. The screened compost is stored for about 30 days in heaps about 2 m wide × 1.5 m high and up to 20m long to ensure stabilisation before sale.

3.4.6.1 Stabilisation in Open Windrows

The stabilisation is carried out in open windrows provided over stone paved or cement concrete paved ground. These windrows are turned every five days to ensure aerobic decomposition. Various types of equipment such as front-end loaders/windrows shifters are used for turning of windrows. At the end of the three to four weeks, the material is known as green or fresh compost wherein the cellulose has not been fully stabilised. It is hence stored in large sized windrows for one to two months either at the plant or the farms. At the end of the storage period, it is known as ripe compost. It may be sometimes subjected to size reduction to suit the requirements.

FIGURE 3.4 Windrow composting. Photograph by the author

FIGURE 3.5 Final Compost product being bagged. Photograph by the author

TABLE 3.18
Analysis of Bio Compost Sample from Site

Parameter	ppm for Iron to Lead
pH	7.67
Electrical conductivity(dsm^{-1})	2.30
Moisture (%)	20.1
Organic Carbon (%)	15.78
Nitrogen (%)	1.12
Phosphorus (%)	0.75
Potassium (%)	0.25
Iron	4920
Manganese	225.0
Zinc	120.5
Copper	70.5
Cadmium	3.72
Chromium	35.15
Nickel	19.52
Lead	16.13

The composting plant essentially should have a leak-proof floor and should be roofed over. The windrow area should be provided with impermeable base. The base can be of concrete or compacted clay of 0.50 m thick having a permeability coefficient less than 10^{-9} m/s. The base should have a slope of 1–2% and be circled by lined slopes for collection of leachate or surface run off. The space between the compost piles should facilitate free movement of operators. Irrespective of the best of practices, there could be leachates coming out of composting piles occasionally. Hence a small slope towards the drain adjacent to the compost platform should be provided. The drain in turn is to be connected to a treatment system which can treat the leachate and reuse the same for irrigation in the fruits and vegetable garden proposed in the plot or to meet the water requirement in composting piles.

3.4.6.2 Organisms Important in Composting

Aerobic composting is a dynamic system wherein bacteria, actinomycetes, fungi, and other biological forms are actively involved. The relative preponderance of one species over another depends upon the constantly changing food supply, temperature, and substrate conditions.

Facultative and obligate forms of bacteria, actinomycetes, and fungi are most active in this process. In the initial stages, mesophilic forms predominate and thermophilic bacteria and fungi soon take over except in the final stage of composting. Except when the temperature drops, actinomycetes, and fungi are confined to 0.05–0.15 m outer surface layer. If the turning is not carried out frequently the actinomycetes and fungi in these layers register increased growth imparting it typical greyish white colour. Thermophilic actinomycetes and fungi are known to grow well in the range of 318–333 K. Different organisms are known to play predominant role in breaking down different constituents of municipal solid waste. Thermophilic bacteria are mainly responsible for the breakdown of proteins and other readily biodegradable organic matter. Fungi and actinomycetes play an important role in the decomposition of cellulose and lignin. The actinomycetes common in compost are *Streptomyces* sp. and *Micromonospora* sp. the latter being more prevalent. The common fungi in compost are *Thermonomyces* sp., *Penicillium dupontii*, and *Aspergillus fumigatus*. Majority of these organisms responsible for composting are already present in municipal solid waste. Not much information is available regarding the organisms active in anaerobic composting, though many of the organisms responsible for anaerobic decomposition of sewage sludge will be active here also, and differences are expected due to the concentration of nutrients present and the temperature conditions. Temperature and time of exposure needed for destruction of some common parasites and pathogens are given in Table 3.30.

3.4.7 Factors Affecting the Composting Process

Important parameters affecting composting process are moisture, temperature, carbon–nitrogen ratio, aeration, particle size and pH

a. **Moisture**
The moisture tends to occupy the free air space between the particles. Hence, when the moisture content is very high, anaerobic conditions set in. However, the composting mass should have a certain minimum moisture content in it for the organisms to survive. The optimum moisture content is known to be between 50 and 60%. Higher moisture content may be required while composting straw and strong fibrous material which soften the fibre and fills the large pore spaces. Higher moisture content can also be used in mechanically aerated digesters. In anaerobic composting the moisture content used will depend upon the method of handling and whether it is carried out in the open or in closed container.

b. **Temperature**
The aerobic decomposition of a gram mole of glucose releases 2830 kJ energy under controlled conditions, while only about 150 kJ are released when it is decomposed anaerobically. Municipal solid waste is known to have good insulation properties and hence the released heat results in increase in temperature of the decomposing mass. Under properly controlled conditions temperatures are known to rise beyond 343 K in aerobic composting. This increased temperature results in increased rate of biological activity and hence results in faster stabilisation of the material. However, if the temperature rise is very high, due to inactivation of the organisms and enzymes the rate of activity may decrease. During anaerobic composting as the released heat is quite small and as part of it is lost from the surface only a marginal rise in temperature occurs.

The studies carried out have shown that the activity of cellulose enzyme reduces above 343 K and the optimum temperature range for nitrification is 303–323 K beyond which nitrogen loss is known to occur. The temperature range of 323–333 K is thus optimum for nitrification and cellulose degradation. The high temperature also helps in destruction of some common pathogens and parasites. During aerobic composting when the material is turned twice in 12 days ***Entamoeba histolytica*** is killed and the eggs of ***Ascaris lumbricoides*** are killed in 36 days when turned thrice. Many studies carried out have shown that the destruction of these organisms is not ensured under anaerobic conditions. It has been proven that the high temperature and long retention during aerobic composting results in destruction of parasites and pathogens. Thus, if the process is so controlled that the temperature is kept between 323 and 333 K for five to seven days, destruction of pathogens, and parasites can be ensured.

c. Carbon:Nitrogen Ratio

The organisms involved in stabilisation of organic matter utilise about 30 parts of carbon for each part of nitrogen and hence an initial C/N ratio of 30 is most favourable for composting. It has been reported that the optimum value is in the range of 26–31, depending upon other environmental conditions. The C/N ratio considers the available carbon as well as the available nitrogen while the available carbon and nitrogen in the MSW may vary from sample to sample. Whenever the C/N ratio is less than the optimum, carbon sources such as straw, sawdust, paper are added while if the ratio is too high, the sewage sludge, slaughterhouse waste, etc., are added as a source of nitrogen. Generally, a ratio of 3 parts to 4 parts browns to 1-part greens works well, but one do not need to be exact about it. In absence of a good mix of brown and green materials, the compost pile may not heat up, may take forever to break down, and start to stink. These issues can usually be remedied easily by adjusting the ratio. If the compost pile is not heating up, then addition of more green material to the compost is required. If there is foul smell, addition of more browns helps. Browns for the compost pile are fallen leaves, sawdust, corn stalks, paper, dry lint, corrugated boxes, etc. Greens include fruit and vegetable wastes, grass, egg shells, animal manure, and plant trimmings.

d. Aeration

It is necessary to ensure that oxygen is supplied throughout the mass and aerobic activity is maintained. During the decomposition, the oxygen gets depleted and has to be continuously replenished. This can be achieved either by turning of windrows or by supplying compressed air. During the turning, it is necessary to bring inner mass to the outer surface and to transfer the outer waste to the inner portion. In case of artificial air supply the quantity of air supply is normally maintained at 1–2 m^3/day/kg of volatile solids. Artificial air supply requires enclosing decomposing mass in containers which is quite costly. Henc the decomposition is commonly carried out in open windrows. Some studies have shown that the optimum turning interval which will reduce the cost and simultaneously maintain aerobic conditions is five days.

3.4.8 Alternate Classification of Raw Materials

The feeds can be classified as bulking materials, energy materials, and balanced materials.

3.4.8.1 Bulking Materials

Dry, porous materials that help aerate the compost pile. They are too low in moisture and nutrients to decay quickly on their own.

TABLE 3.19
Properties of Common Raw Materials in a Composting Plant

Bulking Agents	Energy Materials	Balancing Raw Materials
Low moisture	High moisture	Low to medium moisture
High porosity	Low porosity	Medium porosity
Low nitrogen	High nitrogen	Medium nitrogen
	Typical Raw Materials in a Composting Plant	
Bulking Agents	*Energy Materials*	*Balancing Raw Materials*
Wood chips	Grass clippings	Ground and tree shrub trimmings
Sawdust	Animal dung	Animal bedding
Grass hay	(excludes dogs and cats)	Broad leaves
Wheat or paddy straw	Fruit and vegetable waste	
Corn stalks	Garden trimmings	

3.4.8.2 Energy Materials

These provide the nitrogen and high-energy carbon compounds needed for fast microbial growth. If piled without bulking agent, these materials usually are too wet and dense to allow much air into the compost pile. When you open the pile, it will have a foul, "rotten egg" smell.

3.4.8.3 Balancing Materials

Some raw materials contain a balance of energy and bulking agent properties. These materials will compost readily without being blended with other ingredients (Table 3.19).

Generally, a ratio of three parts to four parts browns to one-part greens works well for the C/N ratio, but one do not need to be exact about it. If the ratio is skewed, the compost pile may not heat up, and may become malodorous. If temperature is to be increased, more green material is to be added to the compost. In case the pile starts smelling adding browns can solve the problem.

3.4.9 Control of Composting Process

The composting is normally taken to be complete when the active decomposition stage is over and the C/N ratio is around 20. If the C/N ratio of compost is more than 20, the excess carbon tends to utilise nitrogen in the soil to build cell protoplasm. This results in loss of nitrogen from the soil and is known as robbing of nitrogen in the soil. If on the other hand the C/N ratio is too low the resultant product does not help improve the structure of the soil. It is hence desirable to control the process so that the final C/N ratio is around 20. The composting process should also be so controlled that the temperature of the decomposing mass remains between 323 and 333 K for at least a week. This ensures the destruction of any parasites or pathogens present in the decomposing mass. During the operation of the process, aerobic conditions should be maintained by controlling the aeration so that smell and odour as well as fly problems do not arise. During turning, care should be taken to avoid dust problem. The

TABLE 3.20
Trouble Shooting in Composting

S. No	Reasons	Prediction	Remedy
		Piles Fail to Heat	
1	Materials too dry	Very difficult to squeeze water	Add water
2	Materials too wet (moisture above 60%)	Liquid squeezed out	Add dry amendments and high carbon ingredients like straw, rice hulls, leaves or shredded paper
3	Not enough nitrogen	Large quantity of woody materials	Add high N ingredients
4	Poor structure	Few large particles, not excessively wet	Add bulking agent
5	Small size pile	Pile height less than 1.1 m	Combine piles
6	pH too low	pH < 5.5	Add wood ash
		Temperature Falling	
1	Low oxygen	Temperature declines gradually, then sharply	Turn or aerate the pile
2	Low moisture	Cannot squeeze water from the material	Add water and mix
3	Composting time nearing completion, C/N ratio less than 20:1	Pile does not reheat after turning or watering	Nothing required
		Pile Overheating	
1	Insufficient aeration	Pile is hot	Turn the pile
2	Moderate to low moisture	Pile feels damp	Add water
3	Pile is too large	Pile height greater than 2.5 m	Split the pile or spread them out
		Odours of Ammonia Rotten Eggs or Putrid Odours	
1	Anaerobic condition, material too wet, poor structure, pile compacted, insufficient duration	Odours, low temperature	Turn the pile. If odour persists add sawdust or any other amendment or bulking agent. Then remix and turn again
2	Anaerobic condition; pile too large	High temperature	Reduce the size
3	Odorous raw material	High temperature	Cover the pile with a thin layer of sawdust to act as an air filter. Turn and mix the materials and bulking materials
		High Levels of Flies and Mosquitoes	
1	Flies breeding on the pile	From manure or food at surface	Turn the pile every four to seven days. Cover static piles with 0.10 m of matured compost or with sawdust or leaves. It no cover, remove outer 0.15 m of material before turning and using it in the core of the new pile so that breeding cycle be interrupted.
2	Flies breeding in raw materials	Wet raw materials stored on site for more than 4 days	Ensure prompt handling of raw materials

windrows should be located over impervious surface so that the surface water from the windrows which may contain entrained particulates and pollutants is properly collected and safely disposed of after processing. Such process leachate can also be reused in composting operation. The rejects from the process should be disposed of at properly designed and operated sanitary landfills. The MSW should be diverted to a properly operated sanitary landfill during annual maintenance period as well as during shutdown of the plant. Ambient air quality and odour should be checked regularly. When the composting is carried out by controlling the various factors within the optimum range, proper quality compost will be obtained. The end product compost should meet the standards prescribed by the government (Table 3.21).

3.4.10 Monitoring

1. Tools:
   a. Thermometer
   b. Hand squeezing
2. Moisture content thumb rule
   Every time after turning/aeration, take handful of composting material, close the fist. When the fist is opened if a ball can be formed and it expands slightly, and if palm is seen slightly moist – 50% moisture.
3. Temperature:
   a. Sufficient aeration – Every day temperature at 1 m deep is not exceeding 6 degrees lower than at 0.3m deep.
   b. Composting activity: Place a piece of metal or wood 0.30–0.60 m inside the pile for 15 minutes. Afterwards, the portion of this piece that was in the pile should feel very hot but not too hot to hold and also it should appear moist.

3.4.11 Properties of Compost

The compost prepared from MSW should be black brown or at least black in colour. It should be crumbly in nature with an earthy odour. The pH should be neutral though slightly acidic or alkaline pH within the range of 6.5–7.5 can be tolerated. The compost should neither be completely dry nor it be lumpy and water should not come out of the mass when squeezed. The nitrogen, phosphorus and potassium (NPK) contents should be more than 1% each. The nitrogen should be in the form of nitrates for proper utilisation by the plants. The C/N ratio should be between 15 and 20 (Table 3.21).

3.4.12 Indian Indore Pit Method of Composting

In 1905 Sir Albert Howard worked as an agronomist in India. He did research on farming practices in the Indian subcontinent which was under British rule and it is not the present-day India. The knowledge he gained in his research on farming practice and composting is the foundation for the current knowledge of the art and science of modern composting practices. He developed the Indore method of composting. This

TABLE 3.21
Specifications of Compost

Parameter	Maximum Acceptable Concentration	
	Organic Compost (FCO 2009)	Phosphate Rich Organic Manure (FCO 2013)
Arsenic (ppm)	10	10
Cadmium (ppm)	5	5
Chromium (ppm)	50	50
Copper (ppm)	300	300
Lead (ppm)	100	100
Mercury (ppm)	0.15	0.15
Nickel (ppm)	50	50
Zinc (ppm)	1000	1000
Parameter	**Specifications**	
Bulk density (kg/m^3)	< 1000	< 1600
Moisture (% max)	25	25
p H	6.5-7.5	(1:5 solution) maximum 6.7
Conductivity, maximum	4.0	8.2
Total organic carbon, % by mass to total mass, minimum	12	7.9
Total nitrogen, % by mass to Total mass, minimum	0.8	0.4
Total phosphate as P_2O_5, % by weight, minimum	0.4	10.4
Total potassium, % by weight, minimum	0.4	-
Total NPK, % by mass to Total mass, min	1.5	-
C:N ratio	< 20:1	< 20:1
Colour	Dark brown to black	-
Odour	Absence of odour	
Particle size	Minimum 90% should pass through 4.0 mm IS sieve	Minimum 90% should pass through 4.0 mm IS sieve

Source: CPCB, 2016.

method was based on the practice of returning plant and animal waste including human waste to the soil. He modified the traditional practice to a systematic method of composting now known as the Indore method.

This method involves layering of organic wastes like fallen leaves, pruning, weeds, stalks, fodder leftovers, chaff, and stem heaped as a pile and then moisturizing and turning the heap at intervals for aerobic decomposition to produce good quality compost. Hard materials are crushed before being piled. Hard materials should be less than 10% of the total plant residue. The mixture of different organic material residues ensures efficient decomposition. Green materials are allowed to wilt for two to three days to remove moisture content before stacking. While stacking, each type of material is spread in layers about 0.15 m thick until the heap is about 1.5 m high. If wet materials are heaped, they will stick to each other and will pack closely. The heap is then cut into vertical slices and about 20–25 kg are put under the feet of cattle in the shed as bedding for the night. In the morning, the bedding along with the dung and urine and urine earth is taken to the pits where the composting is done. The pit should be such a way that rain water should not enter the pit. A temporary shed should be put over the pit to prevent rainfall. The size of the pit should be 1.5–2 m wide and 1 m depth and a suitable length near cattle sheds and should have a water source. The material from the cattle shed is spread in layers of 0.10–0.15m. A slurry made of 4.5 kg dung, 3.5 kg urine earth and 4.5 kg of inoculum from a 15-day-old pit is spread on each layer. Enough water is sprayed on the material to get it wet. The pit is filled in this way layer by layer and it should be filled within one week. The material in the pit should not be compacted in any way.

The material in the pit should be turned thrice during the whole period of composting. The first turning of the whole material in the pit is after 15 days of filling the pit. The second turning is after another 15 days and the third turning after another month. During each turning, the material is mixed thoroughly and moistened with water. The process is continued till the compost is ready for use which takes about four to five months.

3.4.13 Indore Heap Method of Composting

In this method a surface level layered pile is made instead of a pit. In places with heavy rainfall the heap method is used. The compost is prepared in heaps above the ground and a shed is provided to keep the material away from getting wet. The heap is usually 2 m × 2 m × 1.5 m with tapered sides. A protective bund is made around the heap to protect from wind so that the heap doesn't get dried by wind. The heap is usually started with a 0.20 m layer of carbonaceous material such as leaves, hay, straw, sawdust, wood chips, and chopped corn stalks. This is covered with 0.10 m of nitrogenous material like fresh grass, weeds, or garden plant residues. Fresh or dry manure or digested sewage sludge is also added. This pattern is repeated until the heap is 1.5 m high and the material is wetted until it is damp. The pile is sometimes covered with soil or hay to retain heat and it is turned at intervals of 6 and 12 weeks. All the materials can be mixed together retaining the proportion of carbonaceous material and nitrogenous material. Shredding is essential and it can be done by moving a rotary mover over several times or any other appropriate mechanical method. When enough nitrogenous material is not available, green manure, or some leguminous crop is grown on the fermenting heap by sowing seeds after the first turning. The green matter is turned in at the second turning. The process takes four months to complete (FAO et al., 2003). The Indore heap method and pit method are now less preferred because they require high amount of manual labour and operational expenses. Several other operational difficulties and environmental issues are also associated with it. Moreover, several new methods and models for composting like bin composting, in-vessel composting, and larvae composting are being promoted in the present-day solid waste management.

3.4.14 Aerobic High Temperature Composting with Inoculum

Inoculum is a starter culture of microorganisms like bacteria and fungi that are added to the compost pile to speed up the process of biodegradation. There are several branded inoculums for composting available in the market. In addition to the accelerated composting the advantage of such inoculums include odour reduction. nutrient enrichment and improved compost quality. They are available in liquid and powder form, the choice depends on the specific application. Effective microorganisms (EM) consist of common and food grade aerobic and anaerobic microorganisms: photosynthetic bacteria, *Lactobacillus*, *Streptomyces*, *Actinomycetes*, yeast, etc. The strains of the microorganisms are commonly available from microbe banks or from the environment.

3.4.15 Weed Composting

Weeds such as parthenium, water hyacinth are composted using EM method. The materials required are 250 g *Trichoderma viride* and *Pleurotus Sajor-caju consortia*, and urea. A thatched shed is erected first. An area of 5 m × 1.50 m is marked out to lay the materials to be composted. The material to be composted is cut to small size of 100 mm in size. One hundred kilograms of cut material is spread over the marked area and 50 g of microbial consortia is sprinkled over this layer. Again 100 kg of weeds are spread as a layer. One kg of urea is sprinkled uniformly over this layer. This process is repeated until the level rises to 1 m. Water is sprinkled to keep the moisture content between 50% and 60%. Thereafter, the surface of the heap is covered with a thin layer of soil. A thorough turning is done after three weeks. The compost is ready in about six weeks.

3.4.16 Compost Enrichment

Compost can be enriched by the following methods. Application of bonemeal, phosphate rock, or superphosphate to increase phosphate content. Wood ash waste can be added to the compost to increase the K content. Addition of N fixing and P solubilising cultures are done. The quality of compost can be further improved by the secondary inoculation of *Azotobacter*, *Azospirillium lipoferum*, *Azospirillium brasilence* (N fixers), and *Bacillus megaterium* or *Pseudomonas* (P solubilisers). These organisms can be added to decomposing material when it is turned after one month. By this time, the temperature of the compost would have stabilised at about 308 K. These additions also help to reduce the composting time considerably.

3.4.17 In-Vessel Composting

If the composting is done within a building, container, or a vessel it is termed in-vessel method. Forced aeration and mechanical turning techniques are adopted here to accelerate the composting process. There are a variety of in-vessel methods which combine the techniques of windrow composting and aerated pile methods. In-vessel methods use different combinations of vessels, aeration devices, and turning mechanisms. Bin composting is the simplest in-vessel method. Bins can eliminate weather problems, contain odours, and provide better temperature control.

3.4.18 Vermi Composting

Degradation of organic wastes by using various species of worms is called vermicomposting. Vermicompost is the excreta of earthworms which is rich in humus and hence a valuable soil amendment. Vermicompost contains all essential plant nutrients and aids in the growth of plants. Vermicomposting is one of the most environmentally beneficial methods of solid waste management. Application of vermicompost improves growth of new shoots and leaves. Earthworms are grown in brick tanks, pits, crates, tanks, concrete rings, or any other containers and are fed with biomass which is the food for earthworms. The use of earthworms in composting process is found to decrease the composting time and the product is also of good quality. Vermicompost improves soil structure and water holding capacity. Earthworms release substances like auxins and cytokinin which are beneficial for plant growth.

3.4.18.1 The Different Species of Earthworms

Earthworms consume the waste or organic material normally more than their body weight each day. More than two thousand five hundred varieties of earthworms are identified throughout the world and India has more than five hundred native species. Different varieties are identified and the diversity varies with the nature of the soil. *Eisenia foetida* (red earthworm), *Eudrilus eugeniae*, and *Perionyx excavatus* are some varieties.

3.4.18.2 Materials Used for Vermicompost

Wastes from agroindustries, hotel refuse, leaf litter, biodegradable portion of urban and rural wastes, crop residues, weed biomass, and vegetable waste.

3.4.18.3 Method of Preparation

Identify worm-inhabited soils by visible earthworm castings. Dissolve 0.5 kg jaggery and 0.5 kg fresh cattle dung in two litres of water and sprinkle on an area 1 m × 1 m surface soil. Cover with straw lumps and cover with an old gunny bag. Keep sprinkling water for about 20–30 days. Native worms will aggregate in the spot and can be collected and used for preparation of vermicompost. Alternatively, African earthworms are found to accelerate composting. Compost pit of any suitable dimension can be constructed. It can be a single pit, two pits, or tank of any sizes with brick and mortar with proper water outlets. Manageable size is 2 m × 0.6 m × 0.3 m. Arrange a water column to do away with ant menace. A four-chamber pit will facilitate easy and continuous movement of earthworms from one chamber with fully composted matter to the one with the pre-processed waste in the chambers. The Vermibed is prepared as follows:

- Layer 1: Broken bricks and pebbles
- Layer 2: Coarse sand
- Layer 3: Garden soil
- Layer 4: 30-day old bio compost (from the windrow composting)
- Layer 5: One day old biodegradable waste
- Layer 6: Fresh biodegradable waste including cow dung, added every day

1–3 kgs of African earthworms are filled into each bin or trough, to eat through one day's supply of biodegradable waste and cow dung and produce vermi compost – which resembles a dark brown moist mud. Every day's production of vermi compost is removed before adding the fresh layer of biodegradable waste. Regular sprinkling of water is done to keep the bed moisture level at 60 percent. To protect the worms from too much sun and rain, sheltered worm culture is advised. It will be black, granular, lightweight, and humus rich. presence of earthworm casting on the top of the bed is an indicator that the compost is ready. Stop watering the bed prior to three days of harvesting so that the compost is dry and the worms can be easily separated. More than 80% of the worms will move to the bottom of the bed. The harvested compost should be heaped in sun so that all the worms move to the cool base. When more than one pit is there, watering should be stopped in the first pit and the worms will move to the next pit automatically where the environment is right. The worms are separated by using sieves or meshes. The earthworms and the thicker material in the sieves are again composted. The smell of the compost is earth like. If there is a bad odour it means bacterial process is not complete. A musty smell indicates the presence of mould which leads to nitrogen loss. Then aeration is to be done or add more fibrous material and keep the heap drier. The compost is sieved and packed.

3.4.18.4 Advantages of Vermicompost

1. Vermicompost is rich in all plant nutrients as can be seen from the sample analysis. It contains other micronutrients, vitamins, enzymes and plant hormones (Table 3.22).
2. It is free flowing, easy to apply, handle, and store and does not have bad odour.
3. It improves soil structure, texture, aeration, water holding capacity, and prevent soil erosion
4. It provides excellent plant growth.
5. It is rich in beneficial microflora such as N fixers, P-solubilisers, and cellulose decomposers.
6. Vermicompost contains earthworm cocoons and increases the population and activity of earthworms in the soil.
7. It neutralises the soil pH.
8. It prevents nutrient losses and increases the absorption efficiency of chemical fertilisers.
9. Vermicompost is free from pathogens, toxic elements, and weed seeds.
10. It minimises the incidence of pest and plant diseases.
11. Vermicompost enhances the decomposition of organic matter in soil.

12. Vermicompost contains valuable vitamins, enzymes, and plant hormones like auxins which regulates virtually every aspect of plant growth and development, and gibberellins which regulate various development processes including stem elongation, germination, flowering, and flower development.

TABLE 3.22
Vermicompost Analysis

Parameter	ppm for Iron to Lead
pH	6.89
Electrical conductivity (dsm^{-1})	1.27
Moisture (%)	37.4
Organic carbon (%)	15.61
Nitrogen (%)	1.40
Phosphorus (%)	0.77
Potassium (%)	1.09
Iron	5009
Manganese	403.0
Zinc	111.0
Copper	70.5
Cadmium	6.19
Chromium	55.95
Nickel	44.31
Lead	15.97

3.4.18.5 Advantages of Vermicomposting

1. Organic wastes can be broken down and fragmented rapidly by earthworms, resulting in a nontoxic stable material with good structure which has high economic value and act as a soil conditioner which enhances plant growth.
2. Vermicompost supplies suitable mineral balance, improves nutrient availability, and act as a complex fertiliser.
3. Vermicomposting involves great reduction in pathogens and is similar to biocompost.
4. Vermicomposting is an efficient method of disposal of degradable solid waste thus reducing environmental problem.
5. Economic value of solid waste is increased and it can be a cottage industry.

3.4.19 Vermiwash

Vermiwash is another product made by using earthworms. It is plant growth regulator which is rich in macronutrients, micronutrients, enzymes, vitamins, and hormones. Vermiwash is a brown coloured odourless liquid bio-fertiliser. It is a pest and disease control agent and is mainly used as a foliar spray. Studies have shown that it has great value in sustainable agriculture. It enhances the properties of the soil and maintains the fertility of the soil. Application of vermiwash improves soil aeration, soil texture, water holding capacity, and improves the macro and micro nutrient content. Vermiwash promotes root development, increases growth rate, improves crop production, and improves uptake of nutrients present in soil. Vermiwash application is shown to increase the quality of fruits. It also serves as a biopesticide and crop tonic. Vermi wash is not only a liquid manure and plant tonic that helps to reduce many plant diseases, but has other benefits too. A mixture of vermi wash (1 L), cow urine (1 L) and water (10 L) acts as bio-pesticides and liquid manure that increases the rate of photosynthesis in crops, the number of microorganisms in the soil, resistance to pests, and diseases – and hence crop yields. It also increases the rate of production of bio compost. The important components present in vermiwash are:

i. Plant growth-promoting hormones auxin and gibberellins
ii. Enzymes – cocktail of protease, amylase, urease, and phosphatase
iii. Nitrogen fixing bacteria such as *Azotobacter*, phosphate solubilising bacteria

FIGURE 3.6 Arrangement for vermiwash preparation. Photo taken by the author

TABLE 3.23
Analysis of Vermiwash

Component	mg/kg (Except for pH and Conductivity)
pH	8.01
Electrical conductivity (dsm^{-1})	5.69
Organic carbon (%)	0.01
Nitrogen (%)	0.02
Phosphorus (%)	0.003
Potassium (%)	0.39
Iron	19.84
Manganese	0.34
Zinc	0.54
Copper	BDL
Cadmium	0.05
Chromium	0.88
Nickel	0.35
Lead	0.45

It can be obtained as a byproduct of vermicompost units or it can be prepared exclusively.

A clay-pot-based system can be used to produce vermi wash, a liquid fertiliser. Small clay pots with small holes are suspended from the roof and filled with a 100–150 mm base layer of gravel (or small pieces of broken bricks), and a 0.40–0.45 m layer of partially decomposed bio compost. The layers are moistened with water and around 2000 earthworms introduced into the pot. Cotton wicks on bamboo poles are placed in the holes so that the resulting liquid can trickle down continuously. The pot is filled with 4–5 L of water every day. Initially a pale coloured liquid (called vermi-waste) trickles out and after 10 days, vermiwash starts forming in the container – and 3–4 L is collected every day in the larger pot below. Multiple units can be setup to collect the required quantity of vermiwash. The collected vermiwash should be stored in a cool and dry place. Vermiwash is to be diluted and can be sprayed to the growing plants. It can be used to control pests, crop diseases and prevent some fungal and bacterial pathogens (Figure 3.6 and Table 3.23).

3.4.20 Larvae Composting (Entomocomposting)

Entomocomposting is a waste management method that uses larvae to breakdown organic materials. Black soldier fly larvae (BSFL) are the larvae used in this process. BSFL produces compost that is rich in nutrients and hence enhances plant growth. BSFL eats up to 200 mg of food waste per day and can break down waste to 50% in two weeks. BSFL is also capable of removing toxic materials including mercury and heavy metals from compost. The immature larvae and pupae can be used as animal feed. Alternative proteins can also be produced using BSFL. Soldier flies are large insects that resemble wasp and are harmless to human and other animals. They don't have mouth or stingers and hence can't hurt. The life of the fly is two days during which they will be flying around, mating, and laying eggs after which they die.

3.4.20.1 Process

Choose a shady spot in the yard with uninterrupted airflow. Select a bin with a sealable lid. Provide an egg laying substrate like a cardboard to attract adult flies. Bury kitchen waste in the compost pile and cover it with 0.05–0.10 m of dry, absorbent material. Cover nitrogen rich food sources with leaves, dry grass, or shredded paper to avoid attracting females that lay eggs.

The soldier fly larvae can break down manure in a matter of days, reducing the smell, and the chance of carried diseases in areas where animal waste is stored. After the manure piles are broken down to component parts, the worms break away, and they can be collected for chicken feed. They are rich source of protein and birds relish it. The pupae shell is a rich source of chitin and hence a good source of nitrogen in compost (Figures 3.7–3.9).

FIGURE 3.7 Larvae in larvae compost vessel. Photo taken by the author

FIGURE 3.8 Preparation of larvae compost. Photo taken by the author

FIGURE 3.9 Black Soldier butterfly. Photo taken by the author

3.4.21 Thumburmuzhi Model of Aerobic Composting

The *Thumburmuzhi model* aerobic composting system uses a layering technique. One unit consists of two bins with the size of each bin being 1.2 m × 1.2 m × 1.2 m, made of bricks, stainless steel, or concrete and the bottom of the bins are placed 0.4 m above ground level to facilitate leachate collection. The bottom is either conical in shape or with a slope. A leachate collection tank is to be provided with recirculation arrangement. The sides are made with removable bars for manual filling, with a spacing of 80 mm for aeration. For forming the compost pile, first a 0.15 m thick layer of dry leaves are spread and over this organic solid household waste of the same thickness is put. Then cow dung or inoculum is spread. This is repeated till the tank is full. After the first bin is almost full, start filling the other bin. Keep the first bin idle for 30 days for composting to be completed. After 30 days break the pile in the first bin and spread it in the yard for curing. It takes 3 to 4 weeks for stabilising the product. One unit of two bins can be designed to process one ton of biodegradable waste in 60 days. (30 kg biodegradable waste per day) The method uses a consortium of natural microbes and other materials to begin the composting process. Along with minimal greenhouse gas emissions and leachates, the process has three other benefits.

1. Ecologically clean:
   a. Kills pathogens: Being an aerobic recycling model (in the presence of oxygen) the temperature in the layered compost bin rises rapidly, to as much as 70-80 degree Celcius (pasteurization temperature), which curtails the spread of pathogens (bacteria and viruses).
   b. Prevents foul smell and environmental pollution: Aerobic degradation prevents putrefaction and so there is no foul smell or environmental pollution.
2. Easy to maintain: The compost bins need very little space; and the compost unit does not need monitoring.
3. Cost effective: The bin is the only one-time expense and a recurring monthly expense is the cost of the bacterial consortium

3.4.22 Greenhouse Gases Emissions in Composting

In the case of aerobic composting, as seen from Equation 3.16, the gases produced are carbon dioxide, water, and ammonia which later gets converted to nitrates. So essentially carbon dioxide is the greenhouse gas that we have to deal with to reduce climate change effects. In anaerobic digestion process, the gases produced are carbon dioxide, methane, ammonia, and hydrogen sulphide as can be seen from Equation.3.18. It is seen that carbon dioxide and methane constitute 99% of the gases produced in anaerobic stabilisation process. If the methane is not recovered to produce energy, it will have a higher impact on climate change. Even when used as biogas and energy is recovered, the carbon dioxide produced is still a greenhouse gas but less problematic than methane.

The amount of carbon dioxide produced in aerobic composting can be illustrated by an example.

Example 3.6

The organic biodegradable portion of a solid waste can be written as $C_{32}H_{48}O_{30}$. When 100 kg of this is stabilised by composting, 42 kg of compost with chemical formula $C_{15}H_{18}O_{12}$ is obtained. Find the oxygen requirement and the carbon dioxide produced.

SOLUTION

Weight of the organic matter =100 kg

TABLE 3.24
(a) Calculation for Example 3.6

Component	Weight kg	Chemical Formula	Molecular weight (rounded)	kmol = Weight/molecular weight
MSW	100	$C_{32}H_{48}O_{30}$	12×32+1×48+16×30= 912	$\frac{100}{912}$ = 0.10965
Compost	42	$C_{15}H_{18}O_{12}$	12×15+1×18+16×12= 390	$\frac{42}{390}$ = 0.1077

TABLE 3.24
(b) Calculation for Example 3.6 (Contd.)

MSW $C_{32}H_{48}O_{30}$				Compost $C_{15}H_{18}O_{12}$		
Element	kmol	Atomic weight (rounded)	Weight kg	Element	kmol	Weight kg
C	0.10965×32 =3.5088	12	3.5088×12 = 42.1056	C	0.1077×15 =1.6155	1.6155×12 =19.386
H	0.10965×48 =5.2632	1	5.2632×1 =5.2632	H	0.1077×18 =1.9386	1.9386×1 = 1.9386
O	0.10965×30 =3.2895	16	3.2895×16 52.632	O	0.1077×12 =1.2924	1.2924×16 =20.678

Amount of Carbon converted to carbon dioxide = 42.1056−19.386 = 22.7196 kg

$$\text{Moles of C} = \frac{22.7196}{12} = 1.8933 \text{ kmol}$$

$$C + O_2 \rightarrow CO_2$$

I kmol C produces 1 kmol CO_2
Carbon dioxide produced =1.8933 kmol

So, for every 100 kg waste composted 1.8933 kmol CO_2 is produced = 1.8933 × 44 = 83.3052 kg
Therefore, carbon dioxide produced for 100 kg waste composted = 83.3052 kg

From Table 3.23
Oxygen in MSW= 3.2895 kmol
O in the compost = 1.2924 kmol
H present in MSW = 5.2632 kmol
H present in compost = 1.9386 kmol
H converted to H_2O during composting = 5.2632–1.9386 = 3.3246 kmol
Total O requirement = O for CO_2+ O for H=1.8933 × 2 + (3.3246) × 0.5 = 5.4489 kmol
O taken from the waste = 3.2895–1.2924 = 1.9971 kmol
O from air = 5.4489–1.9971= 3.4518 kmol
Weight of oxygen = 3.4518 × 16 = 55.2288 kg
Carbon dioxide produced for 100 kg waste composted = 83.34 kg (Answer)
Weight of oxygen required for 100 kg waste = 55.2288 kg (Answer)

For every 100 kg of degradable MSW composted by aerobic process, 83.34 kg of carbon dioxide is generated. If the MSW is not treated, the same will be in dumps where, it undergoes putrefaction which would produce methane which is a more severe greenhouse gas. Methane has 25 times of impact of carbon dioxide on climate. Methane is considered to contribute to one-fifth of man-made global warming. The climate impact of untreated MSW can thus be reduced appreciably by aerobic process which generates only carbon dioxide. During aerobic composting, the compost product will be nearly 25% of the waste and hence part of the carbon content will be used to enrich the soil. Further, providing appropriate green belt around the composting area can reduce the effects of carbon dioxide. The green belt can consist of various types of trees, shrubs, grass, etc. Fruit-bearing trees and plants as well as flowering plants would add value both from the environmental angle and aesthetic appeal. Further, as proven in small towns across South Asia and East Africa, they provide a revenue income too. The details are given in the case studies in the last section of this chapter.

3.4.22.1 Effect of Increase of CO_2 in the Atmosphere

Human activities alone raised the CO_2 content in the atmosphere by 50% in less than 200 years which in turn causes climate change by warming the planet. The pre-industrial levels were around 278 ppm and at present it is 429 ppm, an increase of 151 ppm. The average rise in temperature of the atmosphere in 2024 is 1.55°C above pre-industrial period according to World Meteorological Organisation. The effects of increased CO_2 levels are summarised here.

1. Global warming
   When there is an increase in CO_2 in the atmosphere, more heat is retained and subsequently an increase in global temperature occurs. This is known as global warming.
2. Climate change
   This increase in temperature is causing climate change by altering the existing weather patterns and causes droughts, storms, and heatwaves.

3. Melting Glaciers
   Higher temperatures in the atmosphere causes the melting of glaciers and icecaps. This in turn contributes to rising sea levels and subsequent disruption of coastal life.
4. Acidification of Ocean
   Oceans absorb 75% of the CO_2 and this CO_2 reacts with sea water forming carbonic acid. This increases the acidity of the oceans which can probably harm marine life like corals and planktons.
5. Impact on agriculture
   Climate change is affecting agricultural productivity. Crop yield and distribution are changed. The impact is in both ways. Some areas have reduced crop productivity whereas there can be benefit in some regions due to longer growing seasons. There is farming sector stress also. With rising temperature, it will be difficult for farmers to work in the fields and farm produce will decrease. In farm sector and construction sector loss of working hours reduces family income. Heat stress and heat waves impacts personal income and slows down economic growth of the nation.
6. Disruption in ecosystem and biodiversity
   The biotic system is very sensitive to climate changes. Some species will not be able to adapt to the temperature and climate changes leading to some biodiversity loss.
7. Health Impacts
   Impact on human health by increased heat related illnesses are a direct consequence of climate change. Reports of deaths and illnesses due to rising temperatures are always reported nowadays. The air quality is also worsened. Heat stress can affect kidney, liver, and brain and may cause illness or even death. Heat related data indicates that heat stress affects more on the poor and marginalised, school, and college going students as there is no cool roof policy in all institutions. Women and children are more affected due to cultural and familial norms of dressing and lack of space for them to get protection from heat. Not meeting the power needs because of the increasing demand brings in power cut and this will affect industrial production also.

Expanding tree cover is crucial to combat the rising CO_2 levels. But that in itself is not enough. Afforestation can act as natural carbon sinks by absorbing carbon dioxide, improve local air quality by trapping air borne pollutants, stabilising soil and providing cooling shade, recharges groundwater, etc., which all contribute to some healthier environment. Though some effects of climate change can be offset, the fact remains that a tree in 40 years will be taking 1 ton of carbon dioxide at the rate of only 25 kg of carbon dioxide per year. Land and resource limitation make it impossible to plant as many trees to offset all carbon dioxide emissions. Large scale reforestation will sequester carbon dioxide and releases oxygen. Trees absorb CO_2 through photosynthesis making them a natural carbon sink. In 2016 a research group of 800,000 volunteers in India planted 50 million tree saplings in an effort to regreen the country. It was found that this project has to be replicated 800 times to cancel out the CO_2 created by humans. Global CO_2 levels can be brought down by planting trees as the National Geographic concludes in a study that "if we act now, we could cut down CO_2 emissions by at least 25% and these levels could not have been seen until almost a century ago".

A study by NASA shows that current increase in CO_2 levels have resulted in a significant greening increase over the last 35 years. This increase in leaves on plants and trees is actually equivalent to planting a forest twice the size of the United States. But the faster growth of these trees results in shorter lifespans of the trees and the time of CO_2 absorption is lesser than anticipated. The species that thrive best on CO_2 levels is important while selecting for carbon sequestration. We need not empirically select the vegetation. We can plant different species and let the fittest survive and propagate (CT, 2022).

3.4.23 Anaerobic Digestion

The quantity of gases produced in anaerobic digestion can be computed from stoichiometric equation if the chemical formula of the biodegradable waste is known. The gases formed are methane, carbon dioxide, and ammonia as given by the equation below.

The generalised stoichiometric equation can be written as (Tchobanoglous, 2019)

$$C_aH_bO_cN_d + \left(\frac{4a-b-2c+3d}{4}\right)H_2O \rightarrow \left(\frac{4a+b-2c-3d}{8}\right)$$

$$CH_4 + \left(\frac{4a-b+2c+3d}{8}\right)CO_2 + dNH_3$$

For an organic waste of chemical formula $C_{32}H_{48}O_{30}N$ the quantity of gases produced can be computed as follows.

For I kmol (926 kg) of the waste if total decomposition takes place

$$\text{Methane produced} = \left(\frac{4a+b-2c-3d}{8}\right) = \left(\frac{4\times 32+48-2\times 30-3}{8}\right)$$

$$= 14.125 \text{ kmol}$$

$$CO_2 \text{ produced} = \left(\frac{4a-b+2c+3d}{8}\right) = \left(\frac{4\times 32-48+2\times 30+3}{8}\right)$$

$$= 17.875 \text{ kmol}$$

If decomposition is only 60% then methane produced $=14.125 \times 0.6 = 8.475 = 8.475 \times 16 = 135.6$ kg.

3.5 REFUSE DERIVED FUEL (RDF)

RDF is defined as the product obtained from the combustible fraction other than chlorinated materials of MSW which has been processed to obtain a high calorific value fuel that is used as a solid fuel alone or with coal in coal-fired power plants, cement kilns, steel manufacturing, etc. It can be used for power generation in boilers. RDF can be the main fuel for electricity generation, biomass power plants, or substitute fuel for industrial boilers. The combustible waste is shredded into a smaller, more uniform particle size for burning. The RDF produced can be burned on site or can be transported to offsite for use as fuel in boilers. In such cases the RDF is to be pelletised. For pelletising the organic components of MSW such as plastic and biodegradable part of the waste is shredded and mixed with additives and binders. The binders used depends on the moisture content, type of waste material, desired fuel quality, and cost effectiveness. Some binders are organic binders like tapioca flour, inorganic binders like calcium hydroxide, natural binders like lignosulphonates, etc. This is a very enriched fuel that can be easily stored and transported. Since RDF is prepared after thorough segregation all harmful materials will be removed. The production of RDF has the steps like segregation, size screening, magnetic separation, and shredding. RDF is usually burned in dedicated incineration facilities with provision for air pollution control. The calorific values are nearly 17,000 kJ/kg. RDF systems are smaller in size as a significant segregation has occurred. The control of combustion is more efficient because of the uniform size and burning characteristics of the fuel. It can also be used in cement kilns, pyrolysis plants, and along with coal. The process system has the following steps and can be represented by Figure 3.10.

3.6 WASTE INCINERATION

Waste incineration refers to the burning of solid waste in excess air at high temperatures of about 1073 K and above in a controlled environment in incinerators. Incinerators are furnaces designed to handle waste materials. This process significantly reduces the volume of the waste by up to 95%. The high temperatures are needed to avoid odour due to incomplete combustion and should be low so as not melt glass. Heat energy, inert gases, and moisture are the products of combustion. Excess air is needed to ensure complete combustion, to promote mixing, and turbulence so that air reaches all parts of the incinerator. Air contains 21% of oxygen and 79% nitrogen. Presence of percentage of excess air determines the composition of the flue gases as flue gas will contain the unused oxygen and a higher percentage of nitrogen coming with the oxygen from air. Consequently, the temperature of combustion will be reduced and this way the combustion can be controlled. For minimising the emission of hazardous components like dioxins, furans, and volatile organic compounds (VOC), high temperatures above 1250 K are needed.

The main combustion reactions are

$C(s)+O_2(g)\rightarrow CO_2(g)$	$\Delta H^0_{298} = -393.51\,kJ$
$H_2(g)+\frac{1}{2}O_2(g)\rightarrow H_2O(l)$	$\Delta H^0_{298} = -285.84\,kJ$
$S(s)+O_2(g)\rightarrow SO_2(g)$	$\Delta H^0_{298} = -296.9\,kJ$

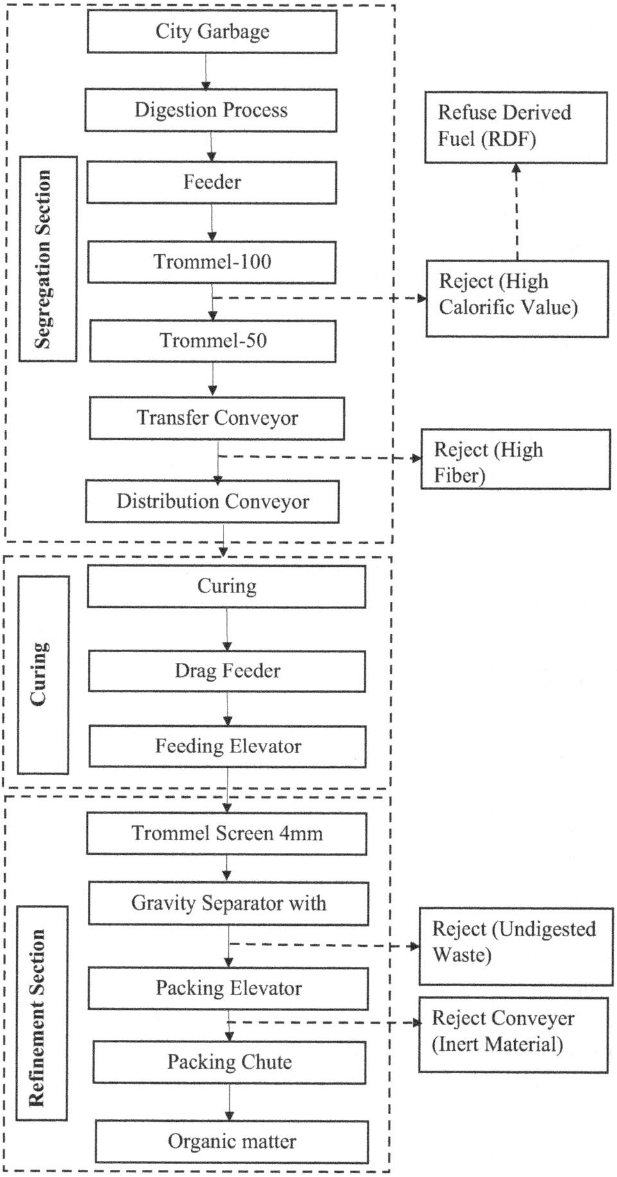

FIGURE 3.10 Schematic diagram of an RDF plant

Incineration is a common method of waste treatment used for various solid waste streams including municipal solid waste, medical waste, and some hazardous wastes. Wastes burned for volume reduction will not need any auxiliary fuel except for startup. Waste incineration can be used for waste to energy conversion because the heat generated during combustion can be used to generate electricity. The process can be Mass-Fired combustion systems where no preprocessing or separation of waste is done. In this type of incinerators all the solid waste is charged into a hopper by cranes. Even non-combustible bulky items will get into the system and the grate system is so designed as to accommodate for the movement of waste materials of all sizes and shapes. The air flow distribution and mixing with the waste for combustion is ensured. Modular incinerators of mass burn type are also in use; these can handle up to 300 tons per day (TPD) of waste whereas larger plants have capacity to burn up to 3000 TPD and they are always part of waste to energy plants.

The incinerator furnaces can be fluidised bed type or Stoker type. Water wall construction is used for the continuously burning type. Here the walls of the combustion chamber are lined with boiler tubes and arranged vertically. Water is circulated which becomes steam using the heat produced by burning. Refractory lining is provided to protect the tubes from excessive heat and mechanical damage. Twin Interchanging fluidised bed combustors with higher efficiency is also available nowadays. These types of incineration have no application in waste treatment of small towns.

3.6.1 Benefits of Solid Waste Incineration

Waste incineration has several advantages though it raises some serious environmental concerns. The environmental issues posed have solutions as appropriate devices can be installed for mitigating the pollution concerns. The use of excess air for incineration also solves the problem to some extent. The consideration in using incineration is where it is appropriate to use this technology and how it can be made part of the waste disposal methods in comparison to other methods. The important advantages are:

i. Volume Reduction: The volume reduction of waste is to the order of 95%.
ii. Energy Recovery: The heat energy from the incinerators can be used to generate power by designing it as waste to energy (WTE) plants.
iii. Waste Destruction: Incineration is an effective method for destroying pathogens and toxic materials. This helps to reduce the public health risks and environmental contamination. The accumulated solid waste from Union Carbide Factory in Bhopal after abandoning the plant after the Bhopal Gas Tragedy is now being incinerated to dispose it off as that is the only viable safe method.
iv. Landfill Diversion: By diverting waste from landfills, incineration can help conserve landfill space and reduce the environmental impact of landfills.

Environmental impact of landfills is very significant and phasing out landfills should be given top priority.

v. Saving on Fossil Fuels: When used for energy generation, waste incineration can partially offset the need for fossil fuels, potentially leading to lower greenhouse gas emissions.

3.6.2 Disadvantages of Solid Waste Incineration

While waste incineration has its benefits, there are also significant drawbacks. Here are some key disadvantages:

i. Air Pollution: As discussed previously, burning waste releases various pollutants into the air such as particulate matter, nitrogen oxides, and harmful toxins like dioxins. These pollutants can contribute to several health problems. But this can be reduced significantly or eliminated totally by using higher temperatures in the range of 1273–1473 K (1000–1200°C) for incineration which reduces harmful byproducts. The exhaust gas from the combustion chamber contains pollutants and needs treatment before discharging to the atmosphere. Scrubbers, filters, and electrostatic precipitators are used to remove particulates, acid gases, and other harmful toxins from the flue gas.

ii. Ash Disposal: The incineration process creates ash residue, which may contain concentrated heavy metals and other toxins. This ash requires careful management and disposal in special landfills.

iii. Greenhouse Gas Emissions: Although incineration can partially reduce reliance on fossil fuels for energy generation, it still releases greenhouse gases during the burning process. (If the waste is not processed then also the aerobic or anaerobic decomposition produces greenhouse gases like carbon dioxide and methane. Methane is a more severe emission as compared to carbon dioxide produced by incineration. These can contribute to climate change and has to be dealt with effectively.)

iv. Incineration tends to discourage Reduction and Recycling: Relying on incineration for waste disposal can discourage efforts to reduce waste generation and prioritise recycling. If there is a convenient way to burn waste for energy, less focus will be given to waste minimisation or recycling.

3.6.2.1 Waste to Energy Plants (WTE Plants)

The main purpose of waste to energy plant is to decrease the amount of waste deposited in landfill and generate energy on a continuous basis by incineration of the solid waste. While composting per se has caught the imagination of many activists and local governments, in actuality it is an unsaleable burden in most of the cities. It is just a myth that compost can be disposed as manure or soil conditioner. That approach is limited to rural areas with large tracts of land. In urban and peri urban areas, it will remain a very romantic idea unless a serious approach to quality management is addressed. Compost has national standards developed and published and if such standards are met with by adding necessary additives to achieve the specifications, there are any number of fertiliser companies which are willing to purchase and market the product in their name. For big cities and urban areas, WTE plants are a more viable option. The WTE plant helps to reduce, reuse, and recycle of the waste through power generation. Part of the energy needs can be met by WTE plants. The incineration of MSW reduces the quantity of waste and produces steam which generate electricity through steam turbines. The incinerator produces ash, gaseous, and particulate emissions and steam. The quantity of energy recovered depends on the nature of the waste. The efficiency of the process depends on the nature of the waste and the characteristics such as moisture content and calorific value. If the moisture content is less auxiliary fuel requirement is only for startup, but if the waste has significant moisture content, then auxiliary fuel requirement reduces the net energy produced from the plant. The combustion process has the following steps. Drying, volatilisation, ignition, elimination of odours, combustion of unburned furnaces gases and carbon particles suspended in the air. The temperature range needed is 1073–1323 K. The minimum temperature required to burn carbonaceous wastes to avoid the release of smoke and prevent emission of furans and dioxins is 1123 K. Maintaining a temperature of 1323K is essential for steam generation and energy recovery and to be assured that all organic compounds are degraded. The flue gas consists of CO_2, H_2O, traces of oxides of nitrogen, sulphur, oxygen, nitrogen, small quantities of hydrogen chloride, mercury, lead, arsenic, cadmium, dioxins, furan, and organic compounds. Flue gas cleaning equipment is to be provided for pollution abatement. The residue includes ash, fly ash, non-combusted organic, and inorganic materials. The products obtained from WTE plants are

i. Power
ii. Compost
iii. Refuse derived fuel (RDF)
iv. Recyclables (Byproduct)
v. Bottom and Flyash (Byproduct)

The power generated depends on the Gross calorific value (GCV) of the MSW. Fresh MSW has a GCV of 5000–5800 kJ/kg. After segregating biodegradables with high moisture content, the remaining waste will have a higher value of GCV of 6300–7100 kJ/kg. The RDF obtained after segregation and composting will be about 35%. This is combustible material with high calorific value and the main components are paper, cardboard, plastic, and wood. The GCV of RDF is about 10,500 kJ/kg. Industrial wastes like bran and rice husk will have higher values of GCV in the range of 12,000–13,000 kJ/kg (Figure 3.11).

For an integrated waste treatment plant operating with 750 TPD of waste with characteristics given in Tables 3.26–3.28 the products obtained are given in Table 3.25.

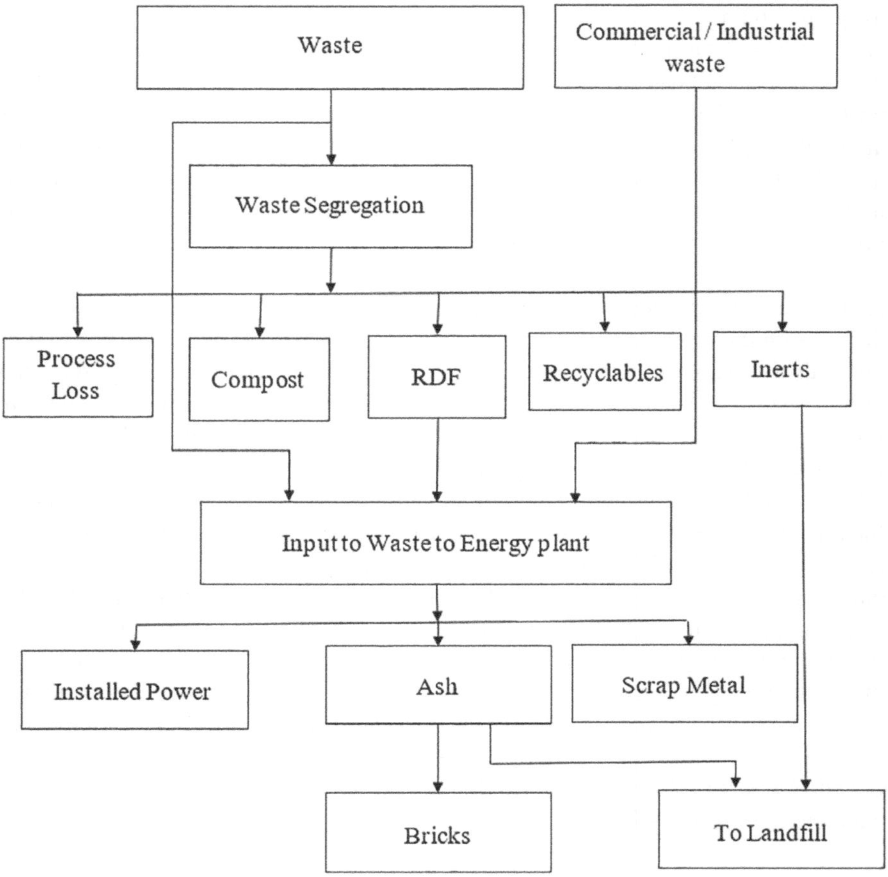

FIGURE 3.11 Schematic diagram of an integrated system with WTE plant

TABLE 3.25
Products from an Integrated MSW plant

S. No	Product	Quantity
1	Power output	15 MW
2	Compost	100 TPD
3	RDF	270 TPD
4	Recyclables	15 TPD
5	Bottom ash + flyash	130 TPD

TABLE 3.26
Typical Physical Composition of Feed to an Integrated Solid Waste Treatment Plant

S. No.	Physical Characteristics	Average Weight %
1	Organic waste	55.0
2	Paper and cardboards	9.0
3	Metal	0.5
4	Glass and ceramics	1.5
5	Plastics	12.0
6	Rags and cotton	6.0
7	Rubber and leather	1.0
8	Garden waste	6.0
9	Others	4.0
10	Inerts	5.0
	Total	100.0

TABLE 3.27
Chemical Composition of Waste

S.No	Parameter	Range
1	Moisture content	38–55% (weight)
2	Carbon content	13.0–33.0% (weight)
3	Nitrogen	1.2–2.5%
4	Phosphorus	0.5–0.9%
5	Potassium	0.4–1.2%
6	Bulk density	341–491 kg/m^3
7	Calorific value	4100–8000 kJ/kg

The chemical composition is as given in Table 3 27. The moisture content typically ranges from 38 to 55%. Calorific value is in the range (4186–7955 kJ/kg)

It is common knowledge that the global accumulation of waste is a long-term environmental threat for future generations. We have not yet found the perfect way to dispose of all of our waste. There are, however, existing technologies that allow us to manage our waste more effectively. One of the best examples of this is responsible solid waste incineration.

TABLE 3.28
Ultimate Analysis of the Waste Feed to WTE Plant

Serial Number	Component	Value
1	Hydrogen as H	4.54%
2	Nitrogen as N	2.52%
3	Oxygen as O	24.02%
5	Sulphur as S	0.27%
6	Carbon as C	33.10%
7	Cadmium as Cd	0.10 mg/kg
	Phosphorus as P	5206 mg/kg
8	Chromium as Cr	17.62 mg/kg
9	Lead as Pb	26.72 mg/kg
10	Arsenic as	0.10 mg/kg
11	Iron as Fe	8516 mg/kg
12	Chloride as Cl	1269 mg/kg
13	Potassium as K	8406 mg/kg
14	Mercury as Hg	0.10 mg/kg

3.7 GASIFICATION OF MSW

Gasification is an age-old technology prevalent from the 19th century essentially to get energy from wood or any cellulosic materials, which were gasified to get fuel for internal combustion engines, furnaces, boilers, etc. Adopting the gasifier technology for reducing the volume of carbonaceous solid waste and generating energy is relatively new. Gasification is the partial combustion to produce a combustible gas by supplying stoichiometrically less amount of air. The gas produced will have a composition CO - 20%, CH_4 - 2%, H_2 - 15%, CO_2 - 10% and the rest N_2. The condensable liquids are also produced along with char and inerts. The calorific value is low when gasified with air as compared to gasification with oxygen.

The gasification reactions are represented by the following equations

$$C(s)+O_2(g)\rightarrow CO_2(g) \qquad \Delta H^0 = -393.51\,kJ$$

$$C(s)+2H_2(g)\rightarrow CH_4(g) \qquad \Delta H^0 = -74.79\,kJ$$

$$C(s)+H_2O(l)\rightarrow CO(g)+H_2(g) \qquad \Delta H^0 = 175.24\,kJ$$

$$CO(g)+H_2O(g)\rightarrow CO_2(g)+H_2(g) \qquad \Delta H^0 = -41.05\,kJ$$

$$C(s)+CO_2(g)\rightarrow 2CO(g) \qquad \Delta H^0 = 172.31\,kJ$$

Gasification technology for MSW has as yet not shown significant success in a commercial scale, though various types of gasifiers are tested. Important types of gasifiers are vertical fixed bed, horizontal fixed bed, and fluidised bed gasifiers. Thermal recovery systems in general have other environmental impacts like gaseous emissions, particulate matter, liquid effluents, and solid residues which are again to be dealt with and requires further control systems. Gasification Technology as such is not yet viable for small towns mainly due to the requirements of consistent and large volumes of waste. The technical challenges of managing the tar and other impurities generated need heavy investments and involve technical complexities.

3.8 PYROLYSIS

Pyrolysis process reduces the volume of MSW and produces useful products pyrolysis oil, pyrolysis gas, and char. The volume of MSW is getting increased day by day and predictions are that it be nearly in the range of 3.5 billion tons by 2050. The yearly rate of increase is estimated to be 6%. Pyrolytic oil is also called tar. Tar has high calorific value and can be used as a fuel. The constituents of tar are hydrocarbons, benzene derivatives, and oxygenated compounds. Char can be used as a fertiliser, soil conditioner, or as adsorbents. Pyrolysis gas consists of carbon monoxide, hydrogen, methane, ethylene, and C_1 - C_3 hydrocarbons. It has a calorific value of 25 - 40 MJ/m^3. The composition depends on the composition of MSW, temperature of pyrolysis, and the type of reactor used. Pyrolysis gas finds used as industrial fuel, crude oil refining, methanol and ammonia production, and as solid fuel substitute. Co-pyrolysis promoted the production of hydrogen. Addition of CO_2 and steam during pyrolysis promotes the reforming reaction and the H_2 and CO content is increased. Pyrolysis is also called destructive distillation. The MSW is finely ground and heated in an oxygen-free atmosphere. The reactions are thermal cracking and condensation reactions and they are endothermic. Pyrolysis was used extensively for wood and now it is being extended to MSW. The pyrolysis reaction for wood is taken as a representative chemical reaction:

$$3(C_6H_{10}O_5) \rightarrow 8H_2O + C_6H_8O + 2\ CO + 2\ CO_2 + CH_4 + H_2 + 7C$$

Pyrolytic oil is represented by C_6H_8O. The oil produced in one plant that was commissioned was 52% and hence the plant had to be shut down as the calorific value is reduced and the product was not saleable. One problem encountered is the inconsistency in the composition and nature of the MSW that is used for the process. Much preprocessing is required before sending to the pyrolysis reactor. The steps involved are shredding, magnetic separation, Air classification, trommeling, and drying to get very fine particles of organic materials.

3.9 SANITARY LANDFILLS

Landfilling or open dumping has been the primary waste disposal method adopted worldwide for centuries. With lesser land availability for dumping the waste and the bad environmental impacts around the waste dumping areas, it became essential that waste dumping or land filling should be scientific. Sanitary landfill is a product of wide thinking in that direction. Sanitary landfills are designed scrupulously to ensure that leachate generation or methane gas emission from such landfills are brought to a minimum and contained scientifically. There are reports saying that the ratio of uncontrolled dumpsites to sanitary landfills is 10:1 which is very alarming. Accordingly, sanitary landfills (SLF) have become a significant part of integrated solid waste management system. Though SLF finds the bottom-most position in the waste management hierarchy the relevance of landfills is here to stay until a stage has arrived at where the zero waste-zero landfill concept is emerging as a reality.

A landfill site also known as dumpsite or dumping yard was used for the disposal of all the waste materials. It is the oldest and most common form of waste disposal. In the past, the waste was simply left in piles or heaps or thrown into low-lying areas of water bodies quite unscientifically. When one dump got filled up it was left as such and new dumping started in another location. Landfills existed even 5000 years ago. There is archaeological evidence that landfills existed in 3000 BCE. In Crete waste was deposited in pits and when full, they were covered with soil. A municipal landfill operated 2500 years ago in the outskirts of Athens. The citizens used to transport their waste to the site to protect their city from the smell of putrefying waste.

When the waste is dumped in open areas unscientifically, it can cause environmental problems like water, soil, air, and noise pollution. It affects the people living in the nearby areas and the ecosystem. It also leads to a lot of health problems that can result in spreading of waterborne diseases and pandemics. Emission of methane gas is one of the most significant environmental concerns regarding landfills. When the biodegradable part of the municipal waste gets decomposed in the absence of oxygen, methane gas is generated and escapes into the atmosphere. This gas is the most potent greenhouse gas and contributes heavily to climate change. Carbon dioxide, water vapour, traces of oxygen, nitrogen, hydrogen, and non-methane gases are also produced from landfills. These gases also can contribute to climate change and when released into atmosphere cause air pollution. The landfill gases contain different components according to municipal solid waste sources and operating conditions of the landfill.

Landfills are potential sources of pollution in the environment that can cause water pollution, air pollution, soil pollution, and pollution of the natural environment. Landfills can contaminate surface water and groundwater resources. Municipal solid waste landfills produce leachate that contains various contaminants. The concentration and quantity of leachate depend on the type of waste deposited in landfills. The leachate usually contains toxic elements and heavy metals, persistent organic pollutants (POP), dissolved methane, fatty acids, sulphates, nitrates, nitrites, phosphates, etc. Leachate percolates through soil and enter and contaminate the surface water or groundwater or nearby water streams. The leachate can affect the microbial diversity and the soil health. The vegetation affected by contaminated leachate can become inedible. Due to unscientific dumping the dumpsites become a breeding ground for mosquitoes and hence spread diseases, inviting animals, and emitting unpleasant odour. There is potential for fires due to the production of flammable gases in landfills because of the activity of microbial degradation deep inside the dumps. Subsurface fires in landfills can smoulder for very long periods of time and emit large quantities of gases such as CO, H_2S, and CH_4 the control of which is difficult.

The generation of landfill gases occurs in five stages, namely, hydrolysis/aerobic degradation, hydrolysis, and fermentation, acidogenesis, methanogenesis, and oxidation. Compared to a speedy aerobic degradation in the initial stages, the conditions for anaerobic degradation processes inside the landfill are not ideal, resulting in slow degradation of the waste material. This leads in turn to long-term environmental effects which can last for several decades, spreading of leachate and emission of pollutant gases persist. Conventional open landfills, which are operated without any control of entering moisture, may require 30 years or even more until they are stabilised.

Open dumping was done previously instead of engineered landfills. These crude landfills are lacking in baseliners, gas ventilation systems and leachate collection systems. Open dumping is now referred to as an inappropriate waste disposal method. To avoid all types of unscientific dumping, it is stipulated that land filling shall be restricted to non-biodegradable, inert waste and other waste that are not suitable either for recycling or for biological processing. It is also made mandatory to avoid land filling of mixed or co-mingled waste. The landfilling shall be done in a Sanitary Landfill, a facility designed with protective measures against pollution of groundwater, surface water, fugitive dust, wind-blown litter, bad odour, fire hazard, bird menace, pests or rodents, greenhouse gas emissions, slope instability and erosion. Sanitary Landfills are designed for reducing exposure of environment and people living in the surrounding areas from toxic waste and emission.

3.9.1 Specification of Landfills

The specifications of landfills are based on the waste management rules of the country (CPHEEO, 2016; EPA, 2014).

There are criteria laid down by the Government for site selection, development of facilities at the landfills, criteria for operation of the landfill, criteria for closure of the land fill, criteria for pollution prevention, criteria for water quality monitoring, criteria for air quality monitoring, criteria for plantation at landfill site, criteria for post care of landfill site, criteria for special provisions for hilly areas, etc. The landfill sites shall be selected to make use of nearby waste processing facilities or a waste processing facility should be set up along with the landfill. The landfill site shall be large enough to last for at least 20–25 years and should be done in a phased manner to avoid waterlogging and misuse. It should be in general 100 m away from river, 200 m from a pond, 200 m from Highways, Habitations, Public Parks and water supply wells and 20 km away from Airports or Airbase. The landfill sites are not permitted within the flood plains as recorded for the last 100 years, in the zone of coastal regulation, in wetland, sensitive eco-fragile areas, etc. A buffer zone is to be maintained around the landfill where there are no development activities. There should be transport facilities, inspection facilities, safety measures for workers, etc. The waste is compacted to increase the density by using heavy compactors and the cell should be covered by a minimum of 100 mm of soil every day. Before the rainy season to prevent infiltration, a layer of 400 to 650 mm compacted soil cover is to be provided and proper drainage is to be given to avoid run off to the active cell of the site. After the site is completed a final soil cover of 600 mm with permeability coefficient less than 1×10^{-9} m/s is to be provided as barrier. Proper drainage and a vegetative layer to support plant growth are to be given.

The leachate and surface run-off water are not permitted to mix and hence are to be diverted. The leachate should not be allowed to flow into any type of waterbodies. The landfills will be receiving hazardous materials like waste oils, bleaches, aerosols, polishes, batteries, etc. So, the landfill should have a soil layer nearly 1 m thick and over which a non-permeable lining system of 1.5 mm thick high-density polyethylene (HDPE) or geomembrane or geosynthetic membrane is to be provided to the

TABLE 3.29
Specifications for Drinking Water Quality

Constituent	Concentration ppm (except for pH)
Arsenic	0.01
Cadmium	0.01
Chromium (as Cr 6+)	0.05
Copper	0.05
Cyanide	0.05
Lead	0.05
Mercury	0.001
Nitrate as NO_3	45.0
Iron	0.3
Total hardness as $CaCO_3$	300
Chlorides	250
Dissolved solids	500
Phenolic compounds (as C_6H_5OH)	0.001
Zinc	5.0
Sulphate as (SO_4)	200
pH	6.5-8.5

As per IS10500, 2012.

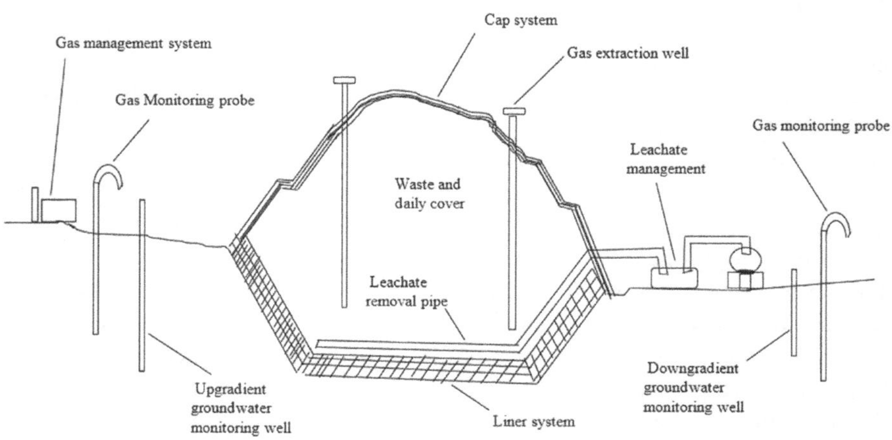

FIGURE 3.12 Schematic representation of a sanitary landfill

base and sidewalls. The groundwater quality in the nearby areas is to be monitored periodically and should meet the criteria as given in Table 3.29.

The gases generated in the landfill are to be collected and for that proper gas collection system is to be installed. The concentration of methane generated shall not exceed 25% of the lower explosive limit. Flaring is to be done if methane generation is high. Plantation should be on the completed landfill. The selection of plants is important as the roots should not go below 300 mm till the stabilisation of the landfill. Locally available plants and nonedible plants which can withstand the climatic conditions are to be selected. The plantation should be so planned that it prevents soil erosion and a green belt also should be provided around the landfill site. Long-term monitoring of the landfills is to done for at least 15 years (Figure 3.12).

3.9.1.1 Zero Waste to Landfill

It is an approach based on the concept of reduction of waste to the minimum so that nothing is left for landfilling. Through this approach the size of the landfill can be reduced and, in the process, need of the landfill can be set aside also. To attain this auspicious goal, there are a lot of obstacles as the path to the zero waste landfill need not be 100% sustainable. Even if the most efficient technologies and systems are adopted there will be some quantity of waste that can only be landfilled. However, zero waste to landfill initiatives are found to be capable of bringing about positive results by reducing quantity of waste and methane emission. Hopefully, it is an achievable target that at least 99% of solid waste can be diverted away from landfills using any or all of the steps in the waste management hierarchy like reduce, reuse, recycle, composting, anaerobic digestion and energy recovery.

3.10 CASE STUDIES

In this section a few field cases of successfully implemented municipal solid waste treatment plants are discussed.

3.10.1 Kattuputhur Town Panchayat, Tiruchirappalli District, Tamil Nadu

Kattuputhur Town Panchayat in Tiruchirappalli District, Tamil Nadu has 3890 houses in 15 wards and a population of about 12,682. A series of measures is being taken to make this municipality a zero waste, garbage-free and green town. To create a garbage-free town, daily collection of biodegradable and non-biodegradable garbage is undertaken. The saleable non-degradable solid wastes are segregated within the town itself before transportation to the Resource Recovery Park (RR Park), name used for waste management centres in any village or town across the state of Tamil Nadu. Even mixed plastic wastes and broken glass have market values and buyers. Basically, waste is just misplaced resource.

3.10.1.1 Salient Features of the Waste Management Facility in Kattuputhur

- ***Non-bio degradable wastes***: The non-degradable wastes like paper, cardboard, plastics, plastic bottles, leather, rubber, metals, glass, and ceramics are collected separately and sold to buyers by the workers who are responsible for waste collection. The non-degradable waste which could not be sold off were transferred to the RR Park for further action. Wastes which could be handled only by specialist agencies like electronic wastes, were kept separately at the RR Park before transferring to the appropriate agencies.

- ***Alternatives to plastic carry bags***: While it is easy to ban or restrict anything, providing the alternatives before the ban is the key to the success of the initiative. The Town Panchayat ensured supply of teak leaves and old newspapers as packing materials in the market before imposing the ban on plastic bags. Yellow coloured cloth bags were traditionally used in Tamil Nadu before the advent of plastic bags. In order to eliminate plastic bags an attractive *rent a bag* scheme was implemented. The bags can be purchased by customers for a very nominal rate per piece but they would be paid the money back when they return the bag to any shop. This kind of incentivisation caught the imagination of the public and hence the scheme became a success in the Panchayat. The scheme was named *Rent a Yellow Bag*. The scheme is supported by Corporate Social Responsibility (CSR) funds from various companies. In order to avoid plastic containers in restaurants and tea stalls, Stainless Steel vessels were supplied by the Town Panchayat to eliminate such plastic containers. Alternatives to plastic items such as earthenware, earthen pot, and wooden spoon were provided to shops and using such items was encouraged.

3.10.1.2 Products from degradable waste and other initiatives

- ***Compost***: The degradable wastes were dealt with initially by setting up a composting facility. Windrow composting is done here. The Town Panchayat and its surroundings have large tracts of agricultural land and hence there was some demand for high value compost.
- ***Vermi composting unit***: The high demand for high-value compost prompted the authorities to set up a Vermicompost unit. The farmers were willing to pay well for the vermicompost as compared to the normal bio compost.
- ***Vermiwash***: Once the vermicomposting unit was a success, the idea of having a new product line too was considered. That was vermiwash which has a good market value and is a highly appreciated and sought after product by the agricultural sector. Foliar Sprays are a part of plant-growing practices. Worm worked soils have burrows formed by the earthworms. Bacteria richly inhabit these burrows, also called as the drilospheres which are part of the soil having earthworm secretions, burrowing and casting. Water passing through these passages washes the nutrients from these burrows to the roots to be absorbed by the plants. This principle is applied in the preparation of vermiwash. Vermiwash is a very good foliar spray.
- ***Lemon Peel Powder***: Segregation of organic wastes for manufacture of value-added products from individual components is an environment friendly approach which improves the economic viability of waste management. In small towns, the efforts both in terms of man-hours and systems for segregation of degradable solid wastes may affect the economics of scale which is the critical factor here. But quite often, it is possible to get some

components segregated at source which can be put to good use. Lemon peels constitute one component which can be used to manufacture value-added products at minimal costs. Citrus peels are known to be a good source of pectin and if available in large quantities, manufacturing pectin is a viable idea. Citrus peels are generally less composting-friendly than other degradable solid wastes due to its properties. For adding to composting piles, the citrus peels need to be cut into smaller pieces and dried so that the composting process maintains the normal pace in the case of open windrows systems. The citrus peels are not recommended in the case of Vermi or larva composting. However, if it is possible to get citrus peels separated from the degradable wastes, manufacturing value-added products is viable option. In households, traditionally, orange peels were used to make marmalades and lemon peels to prepare pickles. Lemon peel pickles are quite popular as a pickle in India. A simple process of solar drying followed by pulverisation of lemon peels can yield a marketable dishwashing powder. In Kattuputhur Town Panchayat, there are several Amman temples where lemon is offered in large quantities and which after the use can be collected as lemon peels separately. Likewise, the small eateries and roadside stalls which sell lime juice make a ready source of segregated lemon peels. Since the quantity available is not sufficient to look at options like extraction of pectin, the idea of shredding into small pieces, making a ground paste followed by desiccation and pulverisation is done. This is just an upgrade of traditional wisdom. Lemon peels had traditionally been used in Indian kitchens to clean oily surfaces. The practical aspects in segregation, cutting into smaller pieces, grinding into a paste and desiccation in sunlight was planned in detail. The sources in the town were advised to keep the lemon peels separated from other degradable wastes. The waste collectors removed them separately and transported them to the Resource Recovery Park. The lemon peel powder, when packed and marketed appropriately did catch the imagination of the citizens making the dish wash powder a commercial success in the Panchayat. The raw lemon peels were found to have moisture up to 85%. The typical characteristics of the final product is as follows. Moisture 8%, acidity 27%, and pH < 4.0. The acidity is caused mainly by the fatty acids like oleic acid and linoleic acid in the lemon peels.

- ***Egg shell Powder***: Another product launched was eggshell powder sold as a natural fertiliser for rose plants. Restaurants and street food vendors were advised to keep the egg shells separate from other wastes so that the same could be collected separately by the waste management workers. The product was marketed with a brand name. The branding, pricing, and marketing strategies were adopted to boost the sales. The egg shell powder was branded as Kattu Rose which means wild rose while indicating the place of origin, Kattuputhur. The product was highly priced.
- All the products like egg shell powder, lemon peel dish wash powder, bio compost, vermi compost, vermi wash and fish waste fertiliser are displayed at the entrance of the Town Panchayat office.
- ***Fish waste Fertiliser***: The fish waste from the market and houses were collected separately. The waste so collected was mixed with banana, jaggery

and water in a tank. The tank was kept closed after 15 days by which time it was ready for application as a liquid fertiliser. The liquid fertiliser was mixed with water 10 times of its volume before application.

- ***Roof top gardening and household vegetable gardens***: The Town Panchayat actively communicated with the citizens, promoting household vegetable cultivation and roof top gardening. Technical support was given to householders in these free of cost by the Town Panchayat officials. Further, the models created by the Panchayat in its office were a sort of an eye opener for the citizen making them aware of the various possibilities.
- ***People's Movement for the Cleanliness of Cities***: The Town Panchayat initiated the People's Movement for the Cleanliness of Cities which is being implemented to encourage the general public to segregate garbage. Saplings are being planted to make it a green town and 3 ponds owned

FIGURE 3.13 Product display at the office. Photo taken by the author

by the municipality are being maintained and saplings are being planted around the ponds. The municipal drains discharging storm water mixed with fugitive sullage were provided with inline treatment to improve the quality of the receiving water bodies. Also, seminars and rallies were held for school students and public to grow saplings, handle garbage and avoid plastic items.

- ***Door Mats and Carpets from Waste clothes***: The equipment for making door mats/carpets from waste clothes were installed in the office and ladies with spare time were engaged to use it. That provides an additional income to them while they have the freedom in the choice of time to work and even the number of mats to be made in a day. The flexibility is a win–win situation for both the operators and the Town Panchayat.
- ***Keeping Pets***: The concept of keeping hens, chicken, duck, fish ponds, rabbits and goats for aiding in the cleaning process is very a thoughtful addition to the waste management drive.
- ***Biogas plant***: A Biogas plant is set up to meet the fuel requirements of the Panchayat office.
- ***Mini-incinerator***: A mini-incinerator is used for the disposal of used napkins and diapers.
- ***Route Maps***: All the route maps for collection of waste are displayed in the Panchayat office. An album covering the news on waste management, photos of visits of VIPs and various awards won are displayed in the Panchayat office.
- ***Product Display***: Product display is done in the Panchayat Office and Plant Office (Figure 3.13).

3.10.2 Ketti Town Panchayat

The project has several innovative features such as

- Processing of faecal sludge from households for production of market quality compost to be used for cultivation of exotic vegetables.
- Recycling of greywater from households for irrigation of exotic vegetables in combination with water storage and farm pond management to have water available during the dry period. Greywater is generated in households from streams without faecal contamination including kitchens and showers.
- It focuses on the women agri-entrepreneurs who traditionally grow vegetables. It enables them to have better quality of crops with competitively priced due to compost application and extended crop season.

Ketti is a Town Panchayat in Nilgiris District in Tamil Nadu. The Town Panchayat has a population of about 34,000. Like in any village or town in the state of Tamil Nadu, Ketti too has a Waste Management Centre which is named Resource Recovery Park. The sanitation system in practice in the Town Panchayat is onsite system in the absence of a sewer system and a Centralized Sewage Treatment Plant. The toilets are mostly flush toilets with a leach pit or a septic tank as the substructure. While there are restrictions on irresponsible dumping of faecal sludge from Septic Tanks and Leach Pits, there was no dedicated place within the town to dispose the Faecal sludge safely. In order to handle the faecal sludge received from the substructures of onsite sanitation toilets, the Town Panchayat initiated setting up a faecal sludge (FS) Treatment Plant in RR Park itself. The RR Park was receiving about 1000 kg of degradable solid wastes from the town. The FS available was on an average 6000 L/week.

Treatment of faecal sludge and sewage in a vertical constructed wetland and co-composting with municipal solid wastes was successfully implemented in Ketti. Vertical Constructed Wetland is a Reed Based Sludge Drying Bed where dewatering and a certain amount of treatment takes place. A vertical flow constructed wetland (VCWL) comprises of a sand surface at the top planted with marsh plants with good root systems and high oxygen secretion capabilities supported by a gravel matrix. The sludge is loaded on the bed and dewatered by percolation and by evapotranspiration through the plants. The root system of the plants maintains the permeability of the sludge layer and sludge can be added continuously. Sludge has to be removed only as and when required while in conventional Sludge Drying Beds, each layer of sludge must be removed when it reaches the desired moisture content, prior to application of the next sludge layer. This has a significant impact on the operational aspect. The long solids retention period favours further mineralisation and pathogen die-off and allows direct reuse of solids in agriculture as a soil conditioner. Percolate quality considerably improves but shall still require a polishing treatment. This system is known as French Constructed Wetland too and used in treating raw sewage directly. The loading rates will be different for sludge and sewage but the design concept remains similar.

The filter and drainage system of a Vertical Flow Constructed Wetland (VCWL) is similar to a Sludge Drying Bed. The plants should be local marshland species that are tolerant to a wide range of environmental conditions (varying humidity, salinity). A freeboard for sludge accumulation shall be provided. Optimal performance has been observed for the loading rate of 250 kg total solids (TS) per m^2 per year as per accepted references. However, a much higher loading rate was possible even in the low temperature area of Ketti. The processes of dewatering and stabilisation of the faecal sludge is achieved in a single treatment stage, unlike any other treatment techniques. Dewatered sludge is used in agriculture as a soil conditioner without further treatment.

The stem of the reeds in the VCWL penetrates through the previous layers of sludge maintains adequate drainage pathways and the plant contributes directly to dewatering through evapotranspiration. The structural facility for a reed bed is similar in construction to an open, under drained sand drying bed. The bottom medium layer provided is 100 mm of 40 mm broken stone: this layer contains the under drain for percolate collection. The layers above this are 100 mm of 20 mm BS and 10 mm BS topped up by 50 mm of 5 mm sand with a top layer of 150 mm with 2 mm sand. The root system of the vegetation absorbs water from the sludge, which is then lost to the atmosphere via evapotranspiration. It is estimated that during the warm growing season this evapotranspiration pathway can account for up to 40% of the liquid applied to the bed. These plants are capable of transmitting oxygen from the leaf to the roots; thus, there are aerobic micro sites (on the root surfaces) in an otherwise anaerobic environment, which can assist in sludge stabilisation and mineralisation.

A periodical harvest of the vegetation is required. The plant stems are cut at a point, which will still be above the top of the sludge layer. This allows the continued transfer of air to the roots and rhizomes. The harvested material is composted. When cleaning of the VCWL is required, the upper portion of the sand layer would be necessary. New sand is then placed to restore the original depth of the media. New plant growth occurs from the roots and rhizomes that are present in the gravel layer. The system has components namely, the faecal sludge collection cum treatment system which essentially is a Vertical Flow Constructed Wetland (VCWL), a Secondary Treatment System, a Horizontal Constructed Wetland (HCWL) for treating the leachate from the VCWL and a dedicated area for co-composting of dried sludge from VCWL with degradable municipal solid wastes.

The faecal sludge was treated in a VCWL after which the treated material was transferred to a windrows compost pile for co-composting with municipal solid wastes. The VCWL had three compartments each with dimensions 5.0 × 3.0 1.5 m with a common underdrain draining the percolate to the HCWL for further treatment. The depth of media was 500 mm with 100 mm deep layers of 40 mm, 20 mm and 10 mm broken stones topped with a 50 mm deep layer of 6 mm broken stone above which coarse sand with a depth of 150 mm. The number of days of operation in Ketti was restricted to 180 days.

The co-composting is done by windrows method in which municipal solid wastes and partially digested faecal sludge from the faecal sludge treatment plant (FSTP) with a moisture content of less than 40% were added in multiple alternate layers like in a club sandwich. Different permutations and combinations of mixing ratios between the two ingredients, dried faecal sludge and degradable municipal solid wastes were tried to arrive at the best combination from the nutrient value perspective. After the pile formation, the composting process took 21 days for compost formation. During this phase, temperature, pH and moisture content were monitored periodically and if there was any deviation from the standards corrective measures were taken immediately. The pile was turned once in a week to ensure the optimum oxygen supply and to avoid the water stagnation and bad odour. After the compost formation, the co-compost is cured (heaped) for 14 days for maturation. The matured co-compost with 17–20% moisture content is sieved to get a particle size of 6–10 mm. The ideal ratio found empirically was one part of partially treated FS to four parts of degradable municipal solid wastes. Once the ideal ratio of mixing was finalised, the programme moved forward to assessing the microbiological quality. The key point in a low maintenance system in a small town was to achieve acceptable microbiological quality at minimum operating cost. Open windrows method of composting is an exothermic reaction generating sufficient heat and temperature for a longer period which in turn achieved good microbiological quality.

The first round of results gave high quantities of helminths but absence of salmonella in the reports. But adjusting the features of the compost piles resulted in complete destruction of helminths too. One takeaway from the trials was avoiding use of inoculum for accelerating the composting process results in better properties of the final product. The sustained high temperatures for longer duration in absence of triggers for accelerating the composting reactions ensured better microbiological quality. The temperatures and the exposure requirements for destruction of various common parasites and pathogens are given in Table 3.30.

The analyses show that with proper control and monitoring and avoiding resorting to exotic inoculums, the national standards can be achieved without difficulty.

The results in the analysis reports over the period of monitoring are as follows (Tables 3.31–3.34).

Indian Standards for heavy metals in compost are given in Table 3.32.

The percolate from the VCWL is fed to a HCWL for treatment before disposal or reuse in the compost pile Table 3.35 shows the analysis reports of VCWL discharges to the HCWL. It can be seen from the reports that the leachate is fit for discharge into land by means of a soak pit or dispersion trench since it meets the Central Pollution Control Board (CPCB) specifications for discharge by those means.

Table 3.36 shows the characteristics of the effluent from the HCWL. Here also, we find the effluent meeting the CPCB standards for disposal to ground by means of a soak pit or dispersion trench (Figures 3.14 and 3.15).

Impacts of co-compost application on soil pH, EC, organic carbon, available nitrogen, phosphorus and potassium, and micro nutrients namely iron, manganese, copper and zinc were assessed. The pH of the soil marginally improved on application of the co-compost. The range before application was 4.3–6.0 which changed to 4.5–6.4.

TABLE 3.30
Temperature and Time of Exposure Needed for Destruction of Common Parasites & Pathogens

S. No	Organisms	Time and Temperature for Destruction
1	*S. typhosa*	No growth beyond 319 K, death in 30 minutes at 328–333 K and 20 minutes at 333 K, destroyed in a short time in compost environment
2	*Salmonella sp.*	In 1 h at 325 K and in 15–20 minutes at 328 K
3	*Shigella sp.*	In 1 h at 328 K
4	*E. coli*	In 1 h at 328 K and in 15–20 minutes at 333 K
5	*E. histolytica* cysts	In few minutes at 318 K and in a few seconds at 328 K
6	*Taenia saginata*	In a few minutes at 328 K
7	*Trichinella spiralis* larvae	Quickly killed at 328 K, instantly at 333 K
8	*Br. abortus* or *Br. suis*	In 3 minutes at 335–336 K and in 1 h at 328 K
9	*Micrococcus pyogenes*	In 10 minutes at 327 K
10	*Streptococcus pyogenes*	In 10 minutes at 327 K
11	*Mycobactercumtuber culosis* var. *hominis*	In 15–20 minutes at 339 K or after momentary heating at 340 K
12	*Corynebacterium diptheriae*	In 45 minutes at 328 K
13	*Necator americanus*	In 50 minutes at 318 K
14	*A. lumbricoides* eggs	In 1 h at 323 K

TABLE 3.31
Co-compost Properties

S. No	Parameter	Unit	Sample	Standards
1	Bulk density	kg/m^3	740	<1000
2	Moisture	%	17.9	15.0–25.0
3	p H	pH units	7.3	
4	Conductivity	dsm	0.5	≤4.0
5	TOC	%	9.0	≥12.0
6	Total nitrogen as N	%	1.53	0.8
7	Total phosphorus as P_2O_5	%	0.46	0.4
8	Total potassium as K_2O	%	0.41	0.4
9	C:N ratio		25.4	25.0–40.0
10	Arsenic	mg/kg	BDL	
11	Cadmium	mg/kg	BDL	
12	Chromium	mg/kg	293.42	
13	Copper	mg/kg	52.53	
14	Lead	mg/kg	20.48	
15	Nickel	mg/kg	65.09	
16	Zn	mg/kg	179.11	
17	Mercury	mg/kg	BDL	
18	Faecal coliform	MPN/g	500	
19	*E. coli*	MPN/g	500	
20	*Salmonella*	MPN/25g	Absent	
21	Helminth eggs	eggs/g	<1	

TABLE 3.32
Limits for Heavy Metals in Compost

S. No	Parameter	Value [in ppm/ (mg/kg)]
1	Arsenic	10
2	Cadmium	5
3	Chromium	50
4	Copper	300
5	Lead	100
6	Mercury	0.15
7	Nickel	50
8	Zinc	1000

TABLE 3.33
Characteristics of Raw Sludge

S. No	Parameter	Unit	Result
1	pH	pH units	6.5
2	Total nitrogen	%	1.74
3	Total phosphorus	%	0.42
4	Total potassium	%	0.054
	Microbiology		
5	Faecal coliform	CFU/g	250,000
6	*E. coli*	CFU/g	140,000
7	*Salmonella*	Per 25g	Absent
8	Helminth eggs	Eggs/g	23

TABLE 3.34
Characteristics of Dried Sludge

S. No	Parameter	Unit	Result 27/11	Result 7/12
1	pH	pH	7.9	7.6
2	Total nitrogen	%	2.3	3.05
3	Total phosphorus	%	0.58	0.11
4	Total potassium	%	0.44	0.50
		Microbiology		
5	Faecal coliform	CFU/g	>1600	>1600
6	*E. coli*	CFU/g	900	>1600
7	*Salmonella*	Per 25g	Absent	Absent
8	Helminth eggs	Eggs/g	30	35

TABLE 3.35
Typical Characteristics of Leachate Input to HCWL

S. No	Parameter	Unit	Value
1	SS	mg/L	144
2	pH	pH Units	6.7
3	BOD	mg/L	82
4	COD	mg/L	152
5	Oil & grease	mg/L	<2
6	Total residual chlorine	mg/L	BDL
7	Ammoniacal nitrogen (as N)	mg/L	24.1
8	Free ammonia (as NH_3)	mg/L	9.8
9	Arsenic	mg/L	BDL
10	Mercury	mg/L	BDL
11	Lead	mg/L	0.28
12	Cadmium	mg/L	BDL
13	Total chromium	mg/L	BDL
14	Copper	mg/L	BDL
15	Nickel	mg/L	0.35

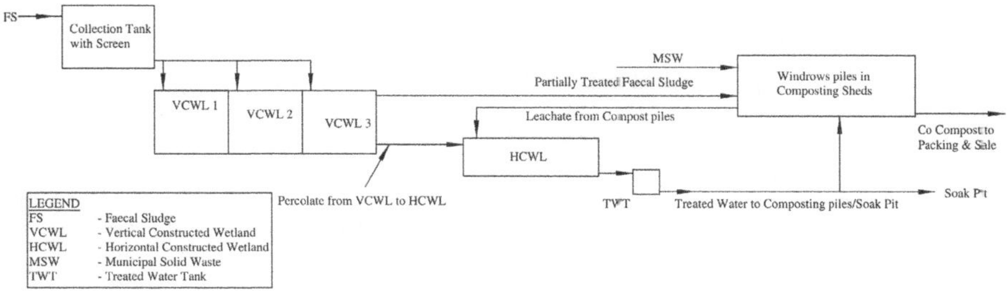

FIGURE 3.14 Process flow diagram for FS and MSW treatment in Ketti

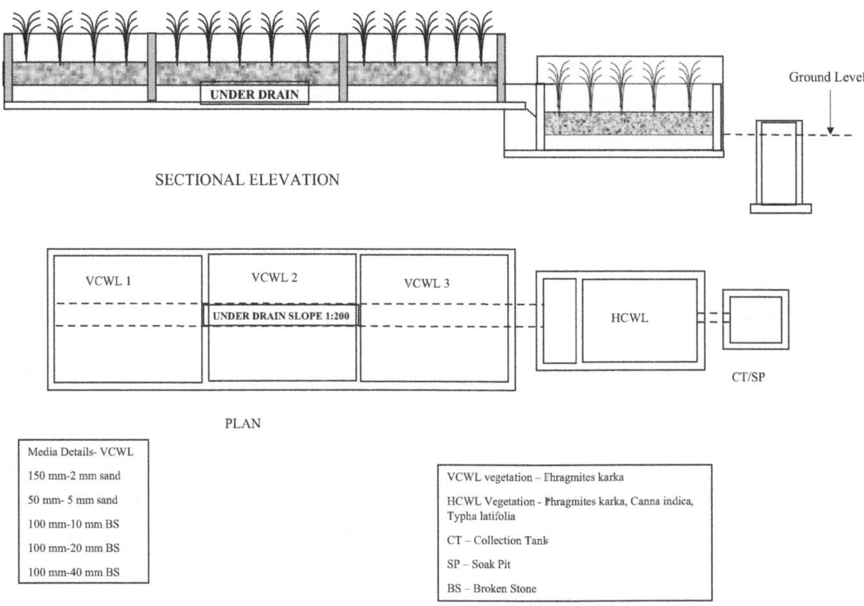

FIGURE 3.15 Layout of wetland for faecal sludge treatment

The co-compost had little impact on the organic carbon content, electrical conductivity or the nutrient content of the soil. However, possibly due to the effect of micro nutrients, the productivity measured in terms of biomass saw a reasonable increase in the case of Beet root (16%) Carrots (19%), Leafy vegetables like parsley (43%). As per the farmers, there are many positive impacts of using co-compost like ease of application and reduction in quantity of chemical fertilisers and reduction in cost of fertilisers, etc. Studies for standardisation of the dose for various vegetable crops were being done by Indian Council for Agricultural Research (ICAR) trying with different doses similar to what had been tried successfully in Balangoda, Sri Lanka.

TABLE 3.36
Typical Characteristics of HCWL Effluent

S. No	Parameter	Unit	Value
1	SS	mg/L	12
2	pH	pH units	7.0
3	BOD	mg/L	16
4	COD	mg/L	80
5	Oil and Grease	mg/L	<1
6	Total residual chlorine	mg/L	BDL
7	Ammoniacal nitrogen (as N)	mg/L	4.5
8	Free ammonia (as NH_3)	mg/L	1.8
9	Arsenic	mg/L	BDL
10	Mercury	mg/L	BDL
11	Lead	mg/L	0.24
12	Cadmium	mg/L	BDL
13	Total chromium	mg/L	BDL
14	Copper	mg/L	BDL
15	Nickel	mg/L	0.32

3.10.3 Adigaratty Town Panchayat RR Park

This is a very unique case in which three different Town Panchayats have a common Resource Recovery Park in one of the Panchayats. The location is in Adigaratty Panchayat. The RR Park is jointly operated by the Town Panchayats of Huligal, Adigaratty, and Jagathaala. This gives some synergy facilitating better returns from the waste management operations. The populations in Huligal, Adigaratty and Jagathaala are about 17,000, 16,000, and 14,000, respectively. The solid wastes generated in each Town Panchayat is about 2500 kg/day.

The RR Park has a single supervisor who is being paid by all the three panchayats. The RR Park has separate collection, sorting and storage areas for the saleable non-degradable wastes. Each Panchayat has its own work force. There are 19 women and three men in the total workforce.

The non-saleable non-degradable wastes are handled separately but stored together for disposal to specialist agencies. There is an FSTP to handle FS from the three panchayats. The FSTP uses Constructed Wetland Technology for the initial processing of the faecal sludge. The entire system is a copy of the one in Ketti except for the capacities. The VCWL for treating the faecal sludge was larger with an area of 100 m^2 instead of the 45 m^2 at Ketti. Likewise, the composting area too was larger at about three times the size of Ketti. The partially treated material from the VCWL is transferred to the Aerobic Windrows piles to be co-composted with degradable solid wastes. The ratio of FS to the degradable wastes is 1:4. The co-compost so prepared is checked for quality and packed in attractive bags for sale. It already had been proved by a study by Indian Council for Agricultural Research (ICAR) that the co-compost manufactured this way is safe and has high nutritional value beneficial for various crops in Nilgiris

FIGURE 3.16 Compost preparation at Adigaratty. Photo taken by the author

FIGURE 3.17 Vertical wetland for treatment of faecal sludge at Adigaratty. Photo taken by the author

FIGURE 3.18 Vertical wetland details at Adigaratty. Photo taken by the author

District. The results obtained in Adigaratty are similar to that of Ketti on all counts including the responses from farmers.

This RR Park sells only co-compost, bio compost and saleable non-degradable wastes. The sale proceeds get divided equally among the three Town Panchayats.

The joint operation results in big savings in

- Land and Infrastructure
- Transportation costs of saleable products going out of the RR Park

- Transportation Costs of incoming materials
- Energy since single connection meets the requirements of all the three local bodies
- Administrative Overheads
- Salaries of the Supervisory Staff
- Single point outlet for Saleable products

This approach had reduced the financial burden of the local bodies in managing wastes and by extension, the system operates at a nominal cost to all the three Panchayats (Figures 3.16–3.18).

3.10.4 Ponnampatty Town Panchayat

Solid Waste Management in small towns can be explained with the help of a case study of a project implemented in a first-grade Town Panchayat Ponnampatty in the state of Tamil Nadu, India. The Ponnampatty Town Panchayat (PTP) has a population of 18,000 and is divided into 15 wards spread over an area of 10.83 km^2. Total number of households is 4120. It has two higher secondary schools, one matriculation school and 9 Panchayat Primary schools, 9 community toilets, two pay and use community toilets. The toilet coverage in the Town Panchayat was always quite good and there are as many as 3000 Septic Tanks in the locality. The faecal sludge (FS) from these Septic Tanks always was a potential environmental hazard. Being a Town Panchayat, the possibility of setting up a Sewage Treatment Plant and a Sewerage system was remote. The solid waste generation is approximately 3000 kg/d. The Panchayath Authorities decided to focus on making the Town Panchayat an eco-friendly, low carbon footprint and waste free panchayat. In Tamil Nadu, the waste management centres are generally called Resource Recovery Parks. The logic was the Waste Management Centre is into activities which yields at least a nominal income for the local bodies since many segregated items are sold at a price. Moreover, there would be production of compost which will yield an additional income to the local body while getting rid of the degradable solid wastes.

The project started off with elimination of plastic bags from the market places and shops. While it is easy to ban or restrict anything, providing the alternatives before the ban is the key to the success of the initiative. The Town Panchayat ensured supply of teak leaves and old newspapers as packing materials in the market before imposing the ban on plastic bags. The attention turned to the RR Park which receives the solid wastes from the town. The RR Park took up multiple activities in a methodical manner.

The first step was a facility for segregation of solid wastes. Non-degradable were classified into various saleable products and sold off at appropriate prices. Most of this activity took place within the town instead of transporting the entire load to the Resource Recovery Park. This included paper, cardboard, plastic bottles, plastic, and metals which had a ready market. The employees who were responsible for collection and segregation could use the sale proceedings for enhancing their income. All those who were supplying the waste were to pay a monthly amount to those who were collecting the waste which also added to the income of those who were collecting and transporting the waste. Transport vehicles and transportation cost is met by the local body. The non-degradable waste which could not be sold off were transferred to the RR Park for further action. Wastes which could be handled only by specialist agencies like electronic wastes were kept separately at the RR Park before transferring to the appropriate agencies. Wastes like discarded clothes were used to produce carpets/door mats using a simple machine. Other solid wastes like construction debris, discarded wood wastes, discarded tyres, and broken ceramic items were used in the entertainment area in the RR Park. The recreation area so created was open to children and families and was quite popular with the citizens of the town. There indeed were wastes which could not be used. While diapers and sanitary napkins are of great calorific value and are ideal fuels in a WTE plant, in absence of any WTE plant in the vicinity and the small quantity to be disposed, an ecofriendly incinerator was set up for disposal of such items.

The degradable wastes were dealt with initially by setting up a composting facility. The Town Panchayat and its surroundings have large tracts of agricultural land and hence there was some demand for high value compost. The degradable are first allowed to pass through a conveyor belt at low speed. Here further segregation is done manually by workers standing on both the sides of the belt. This is followed by size reduction and then windrow composting. A vermicomposting unit too was set up. The farmers were willing to pay the vermicompost twice the price of the normal bio compost. Once the vermicomposting unit was set up, the idea of having a new product line too was considered. That was vermiwash which has a good market value and is a highly appreciated and sought after product by the agricultural sector. Foliar Sprays are a part of plant growing practices. Worm worked soils have burrows formed by the earthworms. Bacteria richly inhabit these burrows, also called as the drilospheres. Water passing through these passages washes the nutrients from these burrows to the roots to be absorbed by the plants. This principle is applied in the preparation of vermiwash. Vermiwash is a very good foliar spray.

FIGURE 3.19 Saplings sale outlet at farm. Photo taken by the author

Segregation of organic wastes for manufacture of value-added products from individual components is an environment friendly approach which improves the economic viability of waste management. In small towns, the efforts both in terms of man hours and systems for segregation of degradable solid wastes may affect the economies of scale which is the critical factor here. But quite often, it is possible to get some components segregated at source which can be put to good use. Lemon peel powder, fish waste manure and egg shell powder were prepared and marketed as described in the case of Kattuputhur.

The RR Park itself has enough land for a limited level of farming. Accordingly, a farm was set up where some of the compost produced was used. The farm yielded as many as 5 saleable products over and above different types of composts and vermiwash. They were fruits, vegetables, flowers, seeds and saplings (Figure 3.19).

3.10.4.1 Faecal Sludge Management

The attention turned to faecal sludge from the 3000 odd Septic Tanks in the Town Panchayat. Instead of handling it as a standalone issue a co-composting unit was set up along with a Low Maintenance Faecal Sludge Treatment Plant. The FS was partially treated in a vertical constructed wetland from which the partially treated sludge was transferred to the co-composting unit using open windrows method. This is a proven method to achieve best microbiological characteristics without diluting the nutrient content of the compost. Treated faecal sludge as a product is of little value since it is at best a soil conditioner only. Compost and co-compost are sold separately based on the preference of the customers.

3.10.4.2 Marketing of Products

The various products being sold by the Town Panchayat are displayed prominently at the entrance of the Panchayat office. The high visibility of the products created a demand for the products. Over and above visibility branding of the products go a long way in increasing the demand and sales. The roof top farm in the Panchayat office which used only the Panchayat compost offered the potential customers a feel of the effectiveness of the compost. The egg shell powder was named Pon Rose a term with multiple meanings and messages. Pon in Tamil means gold/golden. The term Pon also indicates that the product is from Ponnampatty. The attractive name caught the imagination of the potential buyers. Premium pricing was adopted for some products which impressed potential buyers. The egg shell powder with the brand name Pon Rose itself became a big hit with a very high price. Similar product being offered at a tenth of the price elsewhere in Tamil Nadu did not have takers. Premium pricing is an indication of the confidence of the manufacture of the quality assurance angle. The saplings were displayed prominently in the campus of the Panchayat office which too caught the imagination of potential buyers. Beyond that, the Panchayat selling that many saplings was sort of a testimony to the quality of the different types of composts produced by the Panchayat.

3.10.4.3 Encouragement to Rooftop Gardening and Household Vegetable Gardens

The Town Panchayat actively communicated with the citizens promoting household vegetable cultivation and roof top gardening. Technical support was given to householders in these free of cost by the Town Panchayat officials. Further the models created by the Panchayat in its office was a sort of an eye opener for the citizen making them aware of the various possibilities.

3.10.4.4 Economic Viability

Although all the waste management centres in Tamil Nadu are named Resource Recovery Parks, in practice they are all operated in a grant-based model. Effectively, the government, local, district level or state level incurs an operation and maintenance cost (Opex) for waste management, if social cost benefit analysis and intangible benefits are not taken into consideration. The Ponnampatty initiative is a serious attempt to make waste management a viable idea from a commercial point of view too. The basic motto was a one liner – If anything goes out of the RR Park, there should be a corresponding income in cash for that! The RR Park yields 7 different saleable products from waste management namely

- bio compost
- co-compost
- vermi compost
- vermiwash
- fish waste fertiliser
- egg shell powder
- dish wash powder
- door mats

The farm using the compost in the RR Park yields the following items which can be sold for a price.

- fruits
- flowers
- saplings
- seeds
- vegetables

Maintaining the farm helps in reducing the greenhouse gases produced by the composting process. The National Research Centre for Banana (NRCB), Trichy was involved in the assessment of the quality of products in the application they are meant for. This is particularly targeted at the quality control of compost, co-compost, vermi compost, vermi wash and fish waste fertiliser. Optimum usage of each item for specific agricultural produce is being developed.

3.10.4.5 Key to Success-Ponnampatty Approach

- Adoption of only open-source technologies
- Development of new products like egg shell powder for use as fertiliser for rose plants with a registered brand name, Dish Wash Powder with Lemon peels, vermi wash, fish waste-based fertiliser
- Premium pricing of all the products from the RR Centre
- Aggressive marketing of products with visual effects and displays

- Setting up a Faecal Sludge (Septage) Treatment Plant and production of co-compost with degradable solid wastes
- Using sales proceeds of various products to cross subsidise the cost of handling of wastes needing involvement of specialist agencies.

Many Town Panchayats replicated the solid waste management methods adopted here as it was adjudged the best Town Panchayat by the good practices adopted and for improving the quality of life of the people.

3.10.5 Integrated Waste Management Centre, Busia

Busia is a small town located at the western border of Kenya in Busia County and has a population of about 70,000. The Integrated Waste Management Centre was set up in a plot within the campus of Agriculture Training Centre in Busia. The project was funded by FINISH INK, a NPO under FINISH Mondial. The initiative is based on the field application of the concept of Circular Economy. Traditionally, the various types of wastes from a town were considered for treatment and disposal/reuse independent of one another. This project is based on the concept that it is more practical to have a holistic approach to waste management in which several types of wastes taken together is far more sustainable than the traditional way.

The town of Busia has a rural component to it and have the following types of degradable wastes which are processed in the Integrated Waste Management Centre (IWMC).

- Biodegradable wastes from households, market, offices and commercial establishments
- Slaughter house wastes
- Faecal Sludge
- Agricultural wastes

The IWMC have the following facilities

1. *A segregation centre where non-degradable wastes are removed from degradable wastes*
   The processing of the non-degradable wastes is handled by the town management separately. The degradable waste is diverted to the windrows composting facility.
2. *A Faecal Sludge Treatment Plant using Constructed Wetland/Reed Bed system*
   The FSTP is meant to handle 2 truckloads of faecal sludge which translates into 12,000 L/day. The minimum expected solids content in the faecal sludge is 2%. The system consists of a four section vertical constructed wetland (VCWL) planted with *Phragmites* sp. which is fed through a feed tank fitted with stainless steel screens for separation of non-degradable solids. The percolate from the underdrain of VCWL is fed to a Horizontal Constructed Wetland with 6 mm broken stone as media and planted with Phragmites

sp. and some ornamental plants in the periphery for aesthetic appeal The treated effluent is used in the Composting Section. Vector control and odour control are both achieved by the same means-spraying diluted lemon grass oil atop the VCWL just after loading whenever needed.

3. *A Biogas plant*

   A biogas plant of 50 m^3 capacity is set up. Average daily feed rate into the digester are

Type 1 waste from slaughter house	350 kg
Shredded Maize stalks and other wastes	200 kg
Water	550 L

4. *A co composting unit which uses the treated sludge from the above two along with degradable municipal solid wastes and agricultural wastes*

   The capacity of the plant is 3.5 tonnes (of raw material) per day. The system is designed for a mean daily output of 700 kg of co-compost. The compost shed is provided with drains around it which gets treated further in the HCWL or disposed through a soak pit. The compost has been tested in approved laboratories and found to meet the quality standards prevailing in the country and hence safe and effective to be used in agricultural fields. A product development centre where value-added products are manufactured is also there (Figure 3.20).

5. *A demo farm to cultivate fruits, flowers, seeds, saplings and vegetables*

   The demo farm cultivates various produces. Since linked to the Agriculture Training Centre, there is a structured approach to cultivation. A part of the produces from the farm is being sold in the market. The produces include sweet potatoes, water melons, papaya, banana, and various high value leafy vegetables.

6. *The saleable products from the IWMC are*
   - Fruits, Flowers. Saplings, Seeds and Vegetables from the Demo farm
   - Compost
   - Biogas

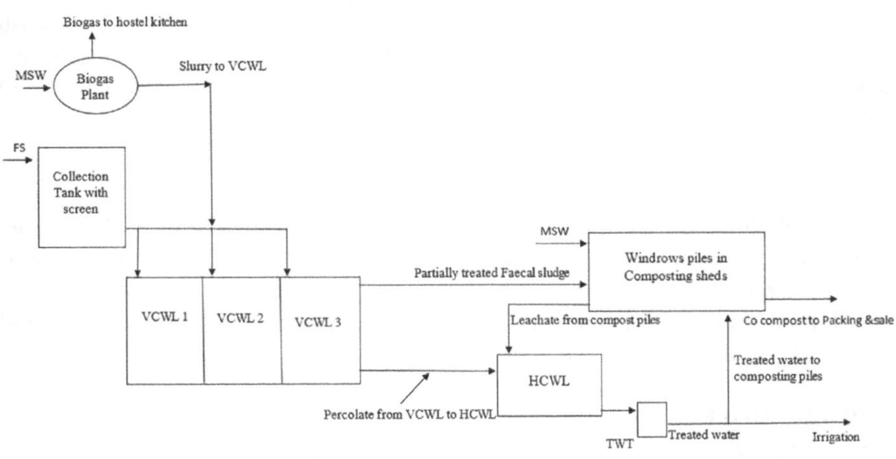

FIGURE 3.20 Process flow sheet for Busia waste treatment plant

7. *Income stream*
   The income for the project includes

   - Tipping fee charged to the operators of the faecal sludge trucks (which is slightly less than the tipping fee charged by the pond-based Sewage Treatment Plant in Busia)
   - Revenue from sale of co-compost and the farm products

3.10.6 The Sanitary Landfill at Dungarpur Town of Rajasthan

Solid waste management of the town was taken up and the following system was set up.

1. Biogas plant
2. Composting unit
3. Sanitary Landfill for non-degradable non-saleable solid wastes

As part of the system a sanitary landfill of capacity 2.5 tons /day was set up. The ultimate objective is leading towards a zero landfill/ zero waste system where the quantity of waste to be land filled is drastically reduced with the present quantity of 2.5 TPD and eliminated over a period of time. The area selected for setting up the Sanitary Land fill (SLF) was adequate for 5 years. Its life can be extended further as and when the quality of waste to be disposed by landfill method will come down. The dimensions used for the landfill unit at the Solid Waste Management Centre were as follows.

Length of landfill	43 m
Width	27 m
Bottom Width	20 m
Depth	3.75 m

After compaction and land filling 0.75 m more above the quantity of total waste that can be landfilled comes to 4562 m^3 for the five years as the rate of 2.5 tons/ day. The system was built accordingly. The periphery of the SLF was provided with safety barriers and a walkway for access and inspection.

3.10.6.1 Lining of Bottom and Side Wall

The bottom and the side walls (slope 1 in 3) are lined with 0.30 m thick amended clay. Before the final compacting of amended clay liner, a leachate collection pit of 1 m × 1 m × 1 m was prepared and the bottom and the side walls lined with amended clay liner as before. The sides of the leachate collection pit too are provided with necessary slopes. Total volume of amended clay required and used is 500 m^3.

3.10.6.2 HDPE Lining

1.5 mm thick (1500 μ) High-Density Poly Ethylene (HDPE) lining was laid over the amended clay layer. 1 m wide HDPE sheets were laid in excess towards all the four sides at the top of the landfill so as to prevent entry of surface run off into it. This part is anchored into the soil. The truck entrance to the SLF was provided with a concrete layer for stability and safety.

3.10.6.3 Leachate Collection Network

After laying HDPE liner over amended clay, a pipe network for leachate collection is provided. The network is made up of HDPE pipes of 20 mm diameter and 0.50 m thick bed of 6 mm broken stone.

3.10.6.4 Preparation of Landfill Site

After excavation the surface of the base of the sanitary landfill was provided with a negative gradient of 1–2% with respect the bottom and side slopes to ensure smooth gravitational flow of leachate and its storage in the leachate collection pit. The leachate collection system is provided with a pump for onward pumping of the leachate to a leachate treatment plant. For levelling the base and the sides of the SLF heavy machinery like tractor and compactor was used.

3.10.6.5 Construction of Liner System

The liner construction is important as it protects the surrounding environment from the impacts of leachate and landfill gas migration. The main components of the liner system include the following.

1. Compacted base
2. Amended clay with porosity 1×10^{-9} m/s to be laid 0.30 m thick over the base and on the side slopes
3. A second layer of amended clay liner installed for the leachate pit 1 m x 1 m x 1 m extending beyond the foot print of the pit.
4. HDPE liner 1.5 mm thick over the compacted clay liner.
5. 1.5 mm thick HDPE liner laid directly above the amended clay liner to act as the primary layer to limit leachate migration and landfill gas migration.
6. A coir cushion layer on the top of the HDPE liner to serve as a protective layer, minimising the risk of damage or puncture during installation of leachate connection network and during the operation of the landfill.
7. 30 mm thick aggregate metal layer on top of the HDPE liner or above the cushion layer to act as a leachate drainage layer. Leachate collection pipes of HDPE 20 mm diameter installed in the drainage layer.

3.10.6.6 HDPE Pipe Network for Leachate Collection

The four outer pipes of the leachate collection network or at least two reaching the collection pit are not be perforated. The perforated pipes surrounded by the bed of aggregates/gravel transport the leachate to the collection pit. A pump is placed within the sump (and within a riser pipe) automatically removes the leachate from the SLF and transport it to the leachate treatment plant at the top. This network is supposed to be a foolproof system to remove the leachate as and when generated and to prevent clogging of leachate in the pipes or at the bottom above the HDPE liner. As we know the leachate may contain acids, hazardous chemicals and other corrosive materials.

If leachate pipe clogs and leachate remain in the SLF, the leachate fluids can build up pressure at the bottom. The liquid pressure becomes the main force for the failure or rupture of the liner.

3.10.6.7 Pumping of Leachate

A riser pipe is installed to pump out leachate from the sump with the help of submersible pump to leachate treatment plant at the top. Submersible pump with automatic level sensor is immersed in the sump.

3.10.6.8 Soil Cover

After each day's filling of wastes or at required intervals, a 100 mm thick soil cover is necessary (the soil is from the excavated lump). Ideally the soil excavated from the site is capable of performing the work. The cover material controls flies, discourages entrance of stray dogs, birds and animals.

3.10.6.9 Gas Vent

Control of gas movement is also an essential function of the cover material. Landfill gases can be either blocked or vented through the cover material. The permeable soil that does not retain much water can serve as a good gas vent. Once permanent cover is laid for each section, corresponding vents will be provided with flares to burn off any combustible gas generated.

3.10.6.10 Fencing

Peripheral fencing is provided at the landfill. It is to limit access to children, trespassers, keep out dogs and other animals. A 2 m fence topped with barbed wire projecting at 45° angle is provided.

3.10.6.11 Control of Surface Water

Surface run off or rain water is diverted from the landfill. Open channels are dug around the top sides of the SLF to drain away the water to ponds and trenches made at low lying areas. If the ponds or trenches get filled pumps are used to prevent flooding.

3.10.6.12 Leachate Treatment Plant

The leachate from the SLF is treated properly for safe disposal or reuse. The treatment system consists of chemical treatment, flocculation, sedimentation, biological treatment in a CWL followed by filtration through a Pressure Sand Filter and an Activated Carbon Filter. For rapid mixing, a non-mechanised Parshall Flume is used instead of a Flash Mixer. This is followed by slow mixing in a Horizontal Baffled flocculator and a sedimentation tank. The clear effluent from the sedimentation tank is treated further in a Horizontal Flow Constructed Wetland followed by filtration through a Pressure Sand Filter and an Activated Carbon Filter before disposal through a soak pit. The sludge settled at the bottom of the sedimentation tank is transferred to the sanitary landfill.

3.10.6.13 Landfill Operation

The compact waste and soil cover constitute a cell. A series of adjoining cells of same height makes up a lift. The completed cover constitutes one or more lifts.

The dimensions of the cell are determined by the volume of the compacted waste. Here two lifts of approximately 1.75–2 m height are formed and about 150 tons of waste i.e. waste segregated in two months forms one cell. After completion of the filling i.e. after 5 years, the cell will have a square shape and its sides have a slope of 20°–30°.

3.10.6.14 Daily Cover

The cover of 0.10 m thick soil is applied to the compacted waste at the end of each operating day or at required intervals.

3.10.6.15 Intermediate Cover

Prior to commencement of monsoon season, an intermediate cover of 0.40–0.60 m thickness of soil shall be placed on the landfill with proper compaction and grading to prevent infiltration. Proper drainage is provided to direct run off away from the active cell of the SLF.

3.10.6.16 Final Cover

After completion of the SLF a final cover shall be provided to minimise infiltration and erosion. The final cover shall have a barrier soil layer of 0.60 m of clay or amended soil with minimum permeability coefficient of 1×10^{-9} m/s. On top of the barrier soil layer shall be a drainage layer of 150 mm. On top of the drainage layer there shall be a vegetative layer of 0.45 m to support natural plant growth and to minimise erosion.

3.10.6.17 Quality of Waste to Be Landfilled

The design is for 2.5 TPD of non-biodegradable waste after removal of reusable and recyclable fractions. It has to be ensured that no biodegradable waste is used for filling the landfill.

3.10.7 Initiatives for Improvement of the Solid Waste Management System

1. Ensuring 100% segregation of waste at source by providing three bins close to generating points like households, markets, institutions, hotels, and hospitals.
2. The concept of decentralising processing of biodegradable wastes at or near the sources through aerobic composting units, biogas plants was initiated.
3. Arrangements for periodic collection of all the non-biodegradable waste and transport to common material recovery facilities where arrangements for further separation of the materials according to their reuse/recyclable quality for value addition are provided. The arrangements include baling machine, shredding machines, extruders and facility for transporting to recyclers and re-processors.
4. Prevention of all types of burning of waste in the landfill, streets, backyards, etc., by strict enforcement of prevailing laws (Figures 3.21–3.24).

FIGURE 3.21 Sanitary landfill. Photo taken by the author

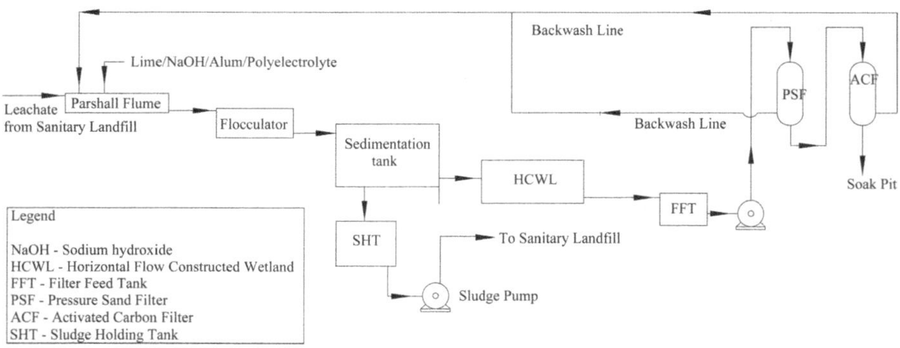

FIGURE 3.22 Leachate treatment plant of sanitary landfill flow diagram

FIGURE 3.23 Leachate treatment plant. Photo taken by the author

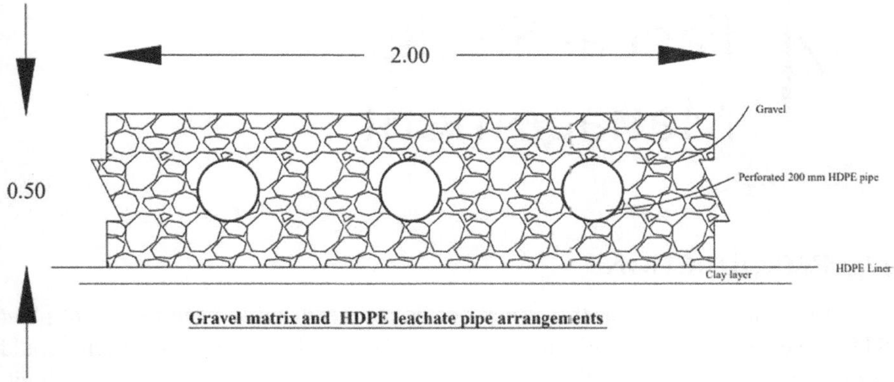

FIGURE 3.24 Details of arrangement for leachate collection

4 Faecal Sludge Management

4.1 INTRODUCTION

In small towns, quite often, there may not be a centralised sewage treatment plant (STP) covering the whole town nor any sewer system. The sanitation system would mainly be in situ systems with one or other substructures like septic tank, leach pit, and ventilated improved pit depending on the type of toilet. In such cases, treatment and disposal of faecal sludge (FS) becomes a challenge. Faecal Sludge Management does not come under either solid or liquid waste management in strictly linguistic terms. FS is often perceived as sewage and sometimes just a part of sanitation. Irrespective of terminologies and classification, FS is part of the wastes found in villages and small towns that do not have centralised sewage treatment facilities and thus depend on in situ facilities. Faecal sludge is often a misunderstood term and in general the terms faecal sludge and excreta are used interchangeably. While excreta are the raw matter discharged from a toilet, the faecal sludge is the material which has undergone a certain amount of biological degradation in one or other means of collection and processing. In an ideal condition, the faecal sludge will be in a fully processed condition and hence in a stabilised form. The type of toilet used has a significant impact on the characteristics of the sludge generated. For urine diverting toilets, there is practically no sludge with just the completely digested material in powder form as the end result. For flush toilets, the main options are septic tanks and Leach Pits for in situ treatment. In the case of a Twin Leach Pit system, the sludge from the pit is already fully digested material which can be used directly as a soil conditioner without any concerns from the health angle. However, they may need to be removed by pumping which in turn warrants dilution of the same by the addition of water. Either way, the material removed need not be considered as sewage but the treatment, disposal, or reuse method starts from dewatering rather than by processing. In the case of septic tanks, when the tank needs to be desludged, the contents could be a mixture of fully digested and stabilised septage, partially digested material, and fresh excreta, which the septic tank might have received that day itself. The case of single pit leach pits too is the same. Poor design and construction of toilets (like the faecal matter from the toilet or effluent from septic tanks or leach pits discharges directly into public drains) can lead to faecal sludge appearing in inappropriate places, causing foul odours, spoiling the aesthetic appeal, contamination of ground water especially shallow water tables, stagnation in pools of dirty water and the consequent spread of water-borne diseases and causing other public health hazards in the area. Even with well-constructed and used leach pit and septic tank, the improper disposal of the faecal sludge emptied from these can create similar problems. In India, operators of hauler trucks that empty these pits and septic

DOI: 10.1201/9781003650737-4

tanks for a fee, without having guidelines and dedicated places for disposal had been dumping faecal sludge near water bodies like lakes, tanks, ponds, canals, and rivers or on barren village lands on the outskirts of larger cities, in the paddy fields and on the roadsides which has caused social protest from these villagers. The need for dedicated locations where the FS can be disposed for treatment and further utilisation cannot be overemphasised. The concept of having dedicated FSTPs has started catching on in recent years.

Traditionally, faecal sludge had been considered a standalone issue to be addressed separately. It is often categorised as a liquid waste and faecal sludge management (FSM) comes under liquid waste management (LWM) with this approach. There had been attempts to take up co-composting of faecal sludge along with municipal solid waste (MSW). However, this approach had some inherent risks regarding the microbiological quality of the final product. Standardisation too was required to have characteristics suitable for field application as manure. FSM as a standalone process often yields a product which functions at best as a soil conditioner than as a manure. In order to demystify the FSM, new approaches were developed in which the FS undergoes a certain amount of treatment before co-composting with degradable MSW at optimum ratios and achieving adequate temperatures for ensuring the ideal characteristics for the final product. Taking this step as a platform further value addition to the co-compost was experimented with successful results.

4.2 COMMON TERMS USED IN FSM

4.2.1 Faecal Sludge (FS)

Faecal sludge consists of all liquid and semiliquid contents of substructures of in situ sanitation systems. It is the material which has undergone a certain amount of biological degradation in one or other means of collection and processing

4.2.2 Black Water

It is the mixture of urine, faeces, and flush water along with anal cleansing water and or dry cleansing material (toilet paper). Black water has all of the pathogens of faeces and all of the nutrients of urine, but in diluted form.

4.2.3 Grey Water

Waste water generated from washing food, clothes, and dishware as well as from bathing. It may contain traces of excreta in some cases. Grey water accounts for approximately 60% of the waste water produced in households with flush toilets. It contains fewer pathogens and dissolved organic matter measured as BOD and COD

4.2.4 Soak Pit

A soak pit, also known as soak away pit or soak well, is a porous-walled chamber that allows treated liquid to slowly soak into the surrounding soil. This is a typical structure used for the disposal of treated effluents.

4.2.5 Septic Tank

A septic tank is the simplest of anaerobic reactors where sedimentation, floatation, liquefaction of solids, and anaerobic digestion of organics take place.

4.2.6 Leach Pit

It is one of the most widely used in in situ sanitation systems. The excreta along with flush water and water from anal cleaning are discharged into the leach pit. The water is distributed uniformly into the surrounding soil by means of the pores or similar means on the wall of the leach pit while the excreta is retained for digestion over a period of time. There are systems with single pits and double pits. Single pits make the toilet unusable once it is filled and hence not recommended unless there is exceptional circumstance. A twin pit system consists of two alternating pits connected to a pour flush toilet. The black water (and in some cases grey water too) is collected in the pits and allowed to slowly infiltrate into the surrounding soil. Over time, the solids are sufficiently dewatered and digested which can be removed for use as soil conditioner.

4.2.7 Septage

Septage refers to the liquid and solid material pumped from a septic tank, cesspool, or other such primary treatment source. Septage has three main components. They are scum, effluent, and sludge.

4.2.8 Desludging

The process of removal of sludge from a tank, pit, or other storage unit. The most suitable method of desludging is by vacuum tankers. Desludging of septic tanks is done when the tank has accumulated too much sludge and scum. A portion of the sludge is retained as seed sludge to ensure that microorganisms responsible for anaerobic digestion remain in the system.

4.3 FAECAL SLUDGE TREATMENT AND DISPOSAL

4.3.1 Characteristics of Faecal Sludge

Where septic tanks or leach pits are connected to a flush toilet, the feed determines the characteristics. The possibilities include all kinds of discharges which include kitchen effluent, laundry effluent, bath water and toilet discharges, excreta alone or a combination of two or three streams mentioned.

In a system with twin leach pits or urine diverting dehydration toilets, the material emptied is almost safe to handle. However, in a septic tank, the situation is somewhat different since the contents could be a mixture of fresh excreta, partially digested excreta, and fully digested sludge. Handling of septic sludge constitutes a serious health risk, since people handling these materials may be exposed directly to pathogens. From human health risk, a basic distinction should be made between sludge, which, upon collection, are still relatively fresh or contains a fair amount of recently deposited excreta and sludge which have been retained in the pits or vaults for months or years and are virtually free of pathogens. In some rare cases, public toilets with just a holding tank for the discharges from the flush toilets are observed and the material removed from such holding tanks are actually putrefied sewage. Special care would be needed both during emptying of toilet pits or vaults by vacuum trucks and during transportation.

Proper inventory has to be made concerning quantities of the faecal sludge that needs to be disposed and the type of sludge that needs to be treated and disposed. A distinction should also be made between fresh and old sludge, and septic tanks containing only toilet water (black water) or those containing toilet and wash water (black and grey water) and, possibly, even rain water. Since the raw faecal sludge often has partially digested or even fresh excreta in it, many FSTPs adopt a system similar to sewage treatment plants (STP) using one or the other aerobic systems for sewage treatment, followed by dewatering of the aerobic sludge obtained from the sewage treatment plants. In cases when the faecal sludge is having a high solids content, the same is often diluted with water first in order to process in the STPs. Prima facie, it looks counterproductive since dewatering should be, in theory, the first step for processing the faecal sludge. In the case of strong FS with high COD the Biochemical Treatment starts anaerobic reactors. In view of these, some among the initiatives in developing countries was to set up and operate faecal sludge treatments adopted the approach focusing on perceiving FS more as sewage while other approach starting from handling digested FS. Hence, the treatment systems should be considered separately for fully digested sludge and partially digested faecal matter mixed with fully digested sludge. This chapter covers the treatment, reuse, value addition, and disposal of faecal sludge as a raw material which is excreta which is digested almost fully with some fresh excreta in it. The chapter on sewage treatment, Chapter 5 of this book details technologies and approach to treatment of raw sewage. Those would apply to the case of faecal sludge which composes of appreciable

TABLE 4.1
Indicative Characteristics of Faecal Sludge

Parameter	High Strength	Low Strength
Characterisation	Highly concentrated FS stored for a few weeks	FS of low concentration stored for years
COD mg/L	20,000–50,000	<15,000
COD/BOD	5:1–10:1	5:1–10:1
NH_4-N mg/l	2–5000	<1000
TS mg/L	≥3.5%	<3%
SS mg/L	≥30,000	7,000 approx.
Helminths no./L	20–60,000	4,000 approx.

Source: Strauss (1996).

quantities of fresh or undigested faecal matter. The typical characteristics of fully digested faecal sludge with both high and low strength are given in Table 4.1. These are just indicative figures and shall vary widely depending on various factors.

4.3.2 Pretreatment of FS

For the treatment of faecal sludge in a treatment facility, it needs to be transported. Dedicated hauler trucks which go by different names like bowser, honey sucker trucks, and gulper are used for this purpose. The system essentially is composed of a tank with pumps capable of pumping the material to be removed through a flexible hose. Pretreatment is a stabilisation step that decreases the odour, the level of pathogens and further decays the sludge before it reaches the location of the treatment plant. Alkali or lime stabilisation is the first step. Lime is added to the liquid to raise the pH to 12 for a minimum of 30 minutes. Approximately 10–12 kg of lime is added for 4000 litres (a typical capacity of a truck) of FS. Lime addition can be at any of the three points like

- To the hauler truck before pumping.
- To the hauler truck while FS is being pumped. (This is the most preferred method since this ensures proper mixing.)
- To the FS storage tank while it is being discharged from the truck.

4.4 TRADITIONAL METHODS OF DISPOSAL OF FS

4.4.1 Land Disposal

It was an old method when the population was thinner and vast tracts of open uninhabited land were available. If land is not used for agricultural activities, then most suitable sites for disposal of faecal sludge waste are isolated areas with clay or rock bottom. In the absence of any of such suitable sites land disposal of faecal sludge is a viable option only if the conditions listed below are satisfied.

a. The site should be away from habitats at least 500 m.
b. The site should be away from drinking water sources. The groundwater table should be at least 10 m deep and surface water bodies should be a minimum 50 m away.
c. The site should be having a low groundwater table and should not be connected to open water. It should be accessible for vehicles like tractor and trailers.
d. It should be within reasonable distance from the point of waste generation.
e. Soil should preferably be clay or rock or other semi-impermeable type.
f. Land should be relatively flat or should be abandoned gravel or clay pits or low-lying areas.
g. The site area should be relatively large.
h. The selected site should have required clearance from concerned regulatory authorities like State Pollution Control Board and Revenue Department.

TABLE 4.2
Advantages and Disadvantages of Land Disposal of FS

Advantages	Disadvantages
1. Low cost involved	1. No reuse of material
2. Simple and easy to construct	2. Difficult to dispose during rainy season
3. No infrastructure and energy requirements	3. Not entirely safe (aerosols)
4. Operation and maintenance cost very low	4. Requirement of large area of land
5. Increase in green cover	5. Leaching from trenches will possibly contaminate ground water and surface water
	6. Support of local people is needed

Land disposal methods are surface application, subsurface application and burial. In surface application, the FS is spread on the site using hauler trucks, specially designed land application vehicles or tank wagons by spray irrigation, ridge and furrow irrigation and overland flow. FS can be applied as a fertiliser or soil conditioner. This method should not be adopted during rainfall or in areas with a shallow water table. For subsurface application the untreated FS is placed just below the soil surface, thus reducing the odour and health risks and also has the benefits of conditioning and fertilising the soil. This method can be used for slopes less than 8% and the soil depth to seasonal water table must be at least 50 centimetres. Major forms of burial include disposal in holding lagoons, trenches, and sanitary landfills. High odour during application is an inherent risk till a final cover is placed on top. Site selection is very important for odour control, vector control and for prevention of pollution of groundwater. Holding lagoons have a depth of maximum 1.80 m and do not allow any soil infiltration. The FS is placed in small incremental lifts of 150 to 300 mm with multiple lagoons loaded in sequential order for drying. In the case of trenches, FS is filled in trenches sequentially in multiple chambers in small lifts of 150–200 mm for optimum drying. Each trench is covered with soil (610 mm) as a final covering and new trenches are opened. The trench can be reused also after drying. Sanitary landfills are not suitable in areas with 900 mm or more rainfall, landfills that do not have leachate prevention and control facilities or those not having underlying rock. A soil cover of at least 150 mm is needed every time of application and a final cover of 600 mm of soil within the final cover after placement of the final lift. Sanitary landfill, though described here, is the last of the options for FSM and should be avoided as far as possible.

4.4.2 Faecal Sludge Reuse without Treatment

Faecal sludge is a good soil conditioner and often used in agriculture. Hence this had been a practical option from the perspective of farmers. The procedure was to have a pit in to which the raw FS is drained out and allowed to dry in the sun before use in the agricultural fields. Farmers were known to pay the hauler truck operators for leaving their truckload of FS in the pits. One of the biggest environmental problems that may result from applying faecal sludge to land is ground water contamination. Direct irrigation with FS too was a practice but it is fraught with many risks, especially the spread of pathogenic organisms on the fields and affecting the crops. Potential for the farm workers getting infected too cannot be overruled. Several bacteria and helminths (worm eggs) may get attached to the plants and infect consumers if the crops are eaten raw and are not thoroughly washed. Accumulation of helminths in the soil can deteriorate the microbiological quality of the soil, resulting in the crops getting affected over a longer period of time.

4.5 TECHNOLOGIES FOR TREATMENT OF FAECAL SLUDGE – PRACTICAL OPTIONS

The options for treatment of FS include

- Thickening
- Settling
- Sedimentation
- Stabilisation ponds
- Natural drying
- Sludge drying bed
- Constructed wetlands
- Co-composting with degradable municipal solid wastes
- Anaerobic digestion

Any of these may be used in tandem with other since the ultimate objective is to have usable solid output and a pollution free liquid. When municipal or domestic wastewater is used in agriculture, the removal or inactivation of excreted pathogens is the principal objective of wastewater treatment. The separation of solids and liquids that make up faecal sludge, is indicated as the 'process of choice' in faecal sludge management. Separation of solids and liquids may be achieved through sedimentation and thickening in ponds or tanks or filtration and drying in sludge drying beds. In case the solid fraction is intended to be used in agriculture, post-treatment is required mainly to meet hygiene requirements. Additional dewatering and drying might be required for landfilling. For the liquid fraction, polishing treatment might be necessary to satisfy criteria for discharge in order to avoid pollution of surface water or ground water. The choice of either sedimentation tanks or ponds, depends on the one hand on the type of sludge to be treated, and on the other hand on the mode of operation envisaged and by the provisions that will be made for handling the mass of solids to periodically be removed from these primary units. In general, solids quantities to be removed from these systems are much smaller in sedimentation/thickening tanks than the mass of solids removed from primary pond systems, with much longer operating cycles (6–12 months instead of a few weeks – a few months). Sludge from septic tanks with a relatively long storage period is biochemically more stable than sludge collected at weekly intervals. Fresh and biochemically active sludge do neither settle nor can they be dewatered, e.g. on sludge drying beds. These sludges have to be mixed with older less fresh sludge at ratios that enable settling. Another possibility is to pre-treat this sludge in a primary anaerobic/sedimentation pond that allows for biochemical stabilisation prior to sedimentation and solids thickening.

While substantial resources have been invested over the past decades into the development of both low and high-cost wastewater technologies, sustainable treatment technologies for faecal sludge still require large inputs of field research, development and testing before they may be propagated as 'state-of-the-art' options.

Properly treated sludge can be reused to reclaim parched land by application as soil conditioner, and as fertiliser in agriculture. Deteriorated land areas which cannot

TABLE 4.3
Matrix Indicating Technologies and Their Application

S. No	Process Technology Type	Process Used for		Products	
		Pretreatment	Post-treatment	Sludge	Liquid Fraction/ Gas
1	Settling Tank	Sludge		Partially dewatered sludge to To step **5/9**	Partially treated water for Treatment steps **3/8**
2	Sedimentation ponds (often a first step in stabilisation pond system)	Sludge		Partially dewatered sludge to To step **5/9**	Partially treated water for Treatment steps **3/8**
3	Stabilisation ponds		Post-treatment of liquids		Treated Water
4	Natural drying		Dewatered sludge	Dried sludge	
5	Sludge drying bed/ sludge dewatering bed	Sludge	Dewatered sludge	Dried sludge	Liquid fraction partly treated water to steps **3/8**
6	Anaerobic digestion Biogas treatment plant	Sludge		Sludge to steps **3/8/10**	Biogas (for use as fuel or to Flare)
7	Vertical flow reed bed for treatment of sludge	Sludge		Mineralised sludge which can be treated further in step **10**	Liquid fraction (partly treated) Needs further treatment before reuse/disposal
8	Horizontal flow reed bed for treatment of liquids	Liquid			Irrigation water. Can be disposed to water bodies/ land
9	Co-composting with organic solid waste			Compost/ fortified manure	
10	Co-composting of partially treated FS with organic solid waste			Compost/ fortified manure	

support the plant vegetation due to lack of nutrients, soil organic matter, low pH and low water holding capacity, can be reclaimed and improved by the application of sludge. Faecal sludge has a pH buffering capacity resulting from an alkalinity that is beneficial in the reclamation of acidic sites like acid mine spoils and acidic coal refuse materials. Faecal sludge contains nutrients in considerable amounts, which leads to fertilisation of soil and organic matters that improve the soil through humic reactions and hence agriculture use of sludge is of top priority (Table 4.3).

Each of the processes is described below.

4.5.1 Settling or Thickening Tank

Settling tanks can be used for partially stabilised faecal sludge (e.g. sludge from septic tanks) and most other sanitation facilities. Settlings tanks are not appropriate for very fresh sludge from public toilets, but may still be suitable if the fresh sludge is diluted with more stabilised sludge. In settling or thickening tanks, the solids accumulate at the bottom and the clarified supernatant usually needs to be further treated. The accumulated sludge is removed periodically through draw-off pipes. Another possibility of sludge removal is manually or by front-loaders after removal of the liquid column and a period of drying. Removed sludge requires further treatment (Figure 4.1).

Settling tanks need sufficient volume for sludge accumulation and sufficient depth of the liquid column (> 1.5 m) to allow good settling. The tanks should be equipped with baffle walls to maintain hydraulic conditions favourable to good settling and to retain floating scum. Design varies depending on the way of sediment removal. If sediment is removed by pumping or hydrostatic pressure, the tank should have a sludge hopper from where the sludge will be drawn off. If sediment is removed manually or by front-loaders the tank will be equipped with an access ramp. At least two parallel tanks will then be needed to assure continuous operation when one of the tanks is due for emptying. The tank size is estimated by choosing the sludge removal interval which varies from two weeks to two months and with the assumed rate of accumulated sludge volume per incoming solids load of 5–9 l/kg total solids (TS) (Table 4.4).

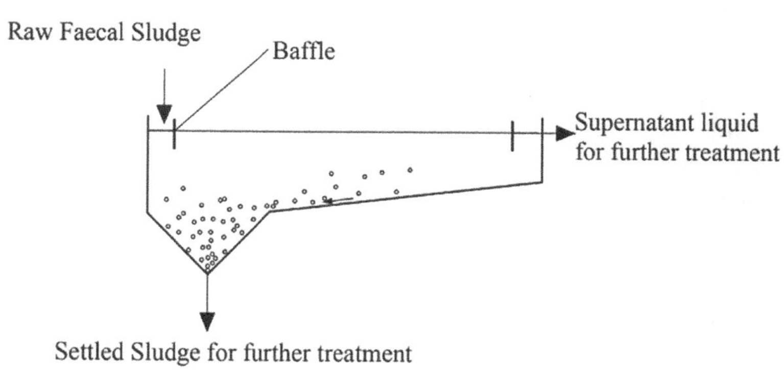

FIGURE 4.1 Settling tank

TABLE 4.4
Advantages and Disadvantages of Settling Tanks

Advantages	Disadvantages
1. Simple and reliable process.	1. Not suitable for fresh faecal sludge.
2. Low requirement of land.	2. Further treatment needed as it only partially dewaters and treats faecal sludge.
3. Partially segregates liquid from solids facilitating further treatment.	

4.5.2 Sedimentation Ponds for Treatment of Sludge

Sedimentation ponds can be used as first FS treatment stage when land availability is not a problem. They can receive fresh FS. Often, they are the first stage in a sequence of stabilisation ponds. The sedimentation ponds use the same principle of sedimentation of solids as settling tanks. Ponds are larger and have longer sediment removal intervals. Due to the high volume and long retention time, they provide a good stabilisation capacity for fresher sludge. The sediment is removed after removal of the liquid column and a period of drying. Both liquid and sediments require further treatment. Sedimentation ponds are designed as anaerobic ponds with a sufficient storage volume for sludge accumulation. Sludge is removed once, twice, or more often per year. At least two parallel ponds are required to assure continuous operation. The organic load of anaerobic ponds is 250–350 g BOD_5/m^3.day; the volume of accumulated sludge per incoming solids load is 0.8-2 l/kg TS (Figure 4.2 and Table 4.5).

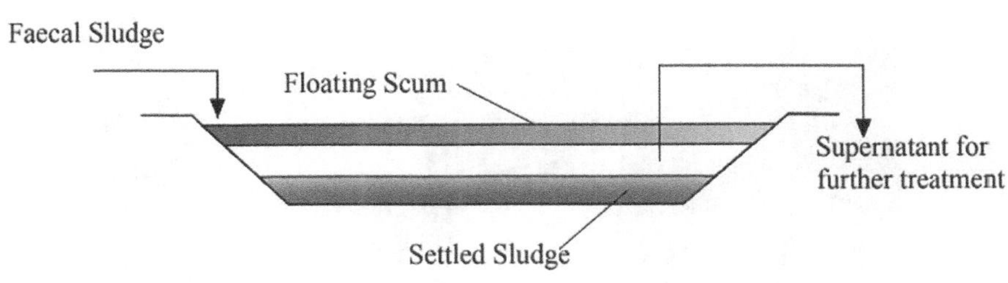

FIGURE 4.2 Sedimentation pond

TABLE 4.5
Advantages and Disadvantages of Sedimentation Ponds

Advantages	Disadvantages
1. Simple operation	1. High land requirements
2. Low construction costs	2. Further treatment is needed
3. Better sedimentation properties compared to settling tanks	
4. Stabilisation capacity is high	

4.5.3 Stabilisation Ponds for Post-Treatment of Liquids

Stabilisation ponds can treat wastewater from sedimentation ponds, settling tanks, etc. Ponds can be used when sufficient land is available. Stabilisation ponds are designed based on the organic loading rates. In anaerobic ponds, an anaerobic environment is created by having a small surface area and high depth. The depth of such ponds is generally in the range 3.0–5.0 m. The retention time is in the range of 20–50 days. The process here breaks down the organic contents and converts them into methane and other gases. Anaerobic ponds are used in tandem with facultative ponds. Facultative Ponds have both aerobic and anaerobic bacteria and depth used is in the range 1.5–2.5 m. The retention time used is less than or equal to 30 days. The BOD removal is highest in a Facultative Pond when compared to other types of ponds at more than 80% (Figures 4.3 and 4.4).

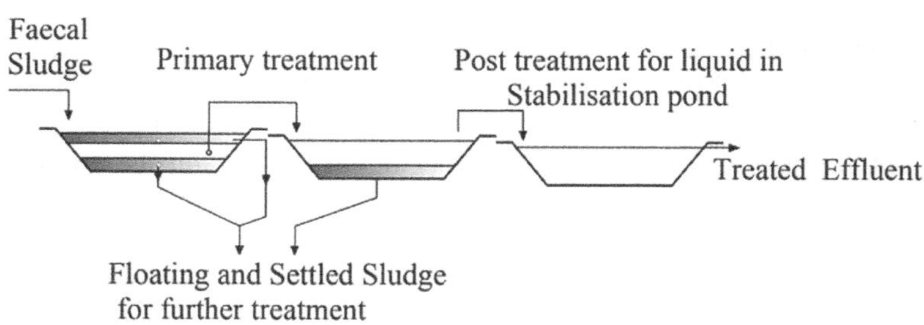

FIGURE 4.3 Stabilisation ponds

FIGURE 4.4 Stabilisation pond. Photo taken by the author

TABLE 4.6
Advantages and Disadvantages of Stabilisation Ponds

Advantages	Disadvantages
1. Simple, well-known and reliable technology. 2. Low Operating and maintenance costs.	1. High land requirements. 2. Possible inhibition of functioning through Ammonia or Ammoniacal Nitrogen in case of very fresh FS.

4.5.4 Natural Drying

Storage on land for over at least 6 months allows natural pathogen die-off in dewatered sludge from settling facilities or drying beds. Further drying of sludge contributes to pathogen die-off and increases the safety of the method. Protection against rain may be required depending on the climatic conditions. Storage and natural drying will be used if the faecal sludge is to be reused in agriculture and if co-composting or constructed wetlands (other processes delivering hygienically safe bio solids) are not favoured (Tables 4.6 and 4.7).

TABLE 4.7
Advantages and Disadvantages of Natural Drying

Advantages	Disadvantages
Low Operation and maintenance costs and simple	High land requirement

4.5.5 Drying Beds for Treatment of Solid Fraction

Drying beds typically consist of a gravel plus sand filter with a drainage system at the bottom. Raw or pre-settled FS is loaded on the bed, and the water is evacuated mainly by percolation through the filter and to some extent by evaporation. The dewatered sludge can be disposed. Further treatment for pathogen removal is necessary if the dried sludge is to be reused. Percolate quality improves through filtration but usually requires polishing treatment. Various designs have been developed for drying of digested sewage sludge. The most frequently used type is the sand drying bed. They can be designed for a loading rate of 150–200 kg total solids (TS) per m^2 an year. Dried sludge can be removed after 7–14 days, depending on climatic conditions.

Drying beds can be used as first treatment stage and as second stage for dewatering of settled sludge removed from settling or thickening tank. Drying beds cannot receive undiluted fresh FS (poor dewatering characteristics, odour emissions). Drying beds can be used for dewatering and drying of septage, septage/public toilet sludge mixtures at volumetric ratios > 2:1 and of primary pond sludges with initial TS content varying from 1.5 to > 7%. Dewatering performance varies with the initial TS and total volatile solids (TVS) content and the applied loads (Figure 4.5 and Table 4.8).

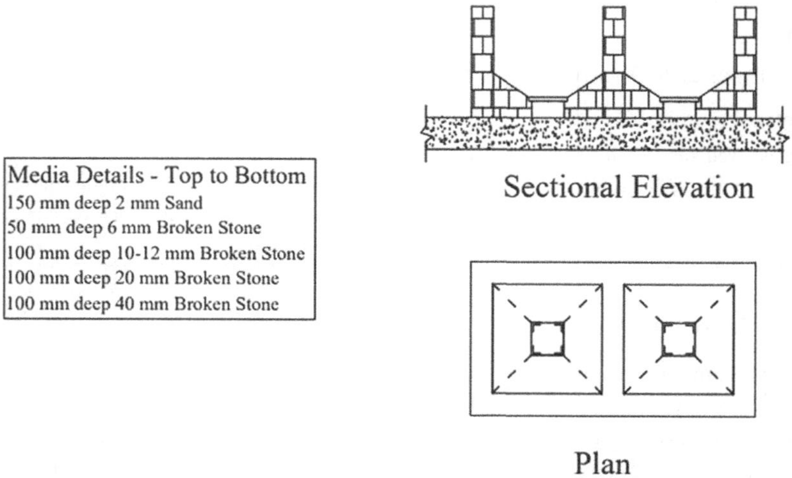

FIGURE 4.5 Drying bed for solid sludge

TABLE 4.8
Advantages and Disadvantages of Drying Beds

Advantages	Disadvantages
1. Low moisture content of dried solids facilitating easy disposal.	1. Solids are not yet hygienically safe (unlike constructed wetlands).
2. Relatively good percolate quality (compared to settling facilities).	2. Low drying propensities in rainy season.
3. Technology is well-known and reliable.	

4.5.6 Constructed Wetlands for Treatment of Sludge/Effluent

Constructed wetlands can be used when the sludge is to be used in agriculture (CPCB, 2003). Two major types of constructed wetlands are used for sludge and effluent treatment according to the direction of the flow. The vertical flow system is suitable for treatment of faecal sludge, whereas the horizontal and vertical flow systems are used for liquid waste treatment. The exact treatment process in a reed bed is still not fully known. It is believed that essentially the process is similar to any other biological treatment system except that various forms of microorganisms such as aerobic, anaerobic, or anoxic may co-exist. It is believed that a significant portion of the microorganisms responsible for treatment are aerobic in nature and they generally find the root structure of reeds and other plants as their habitat. They get the atmospheric oxygen transferred by the reeds to the root structure. Although reed beds were originally developed in industrialised countries, results from Kerala and Tamil Nadu (India) show that the combination of various microorganisms provides effective treatment to complex organic effluents It may be that favourable climatic conditions (high temperature and sunshine) contribute at least partially to the performance.

4.5.6.1 Vertical Flow Treatment System to Treat Faecal Sludge

A vertical flow constructed wetland comprises gravel and sometimes sand and is planted with marsh plants. The sludge is loaded on the bed and dewatered by percolation and by evapotranspiration through the plants. The root system of the plants maintains the permeability of the sludge layer and sludge can be added continuously. Sludge has to be removed only once every few years. The long solids retention period favours further mineralisation and pathogen die-off and allows direct reuse of solids in agriculture. Percolate quality considerably improves, but may still require a polishing treatment. The filter and drainage system of constructed wetlands is similar to a drying bed. The plants should be local marshland species that are tolerant to a wide range of environmental conditions like varying humidity and salinity. A freeboard for sludge accumulation of up to 1 m should be provided. Optimal performance has been observed for the loading rate of 250 kg total solids (TS) per m^2 per year. The sludge accumulation is then approximately 200 mm per year. Post-treatment of liquid effluents from primary treatment assures that the final effluent can be discharged into surface waters with no harm for the environment and public health (Figures 4.6 and 4.7; Table 4.9).

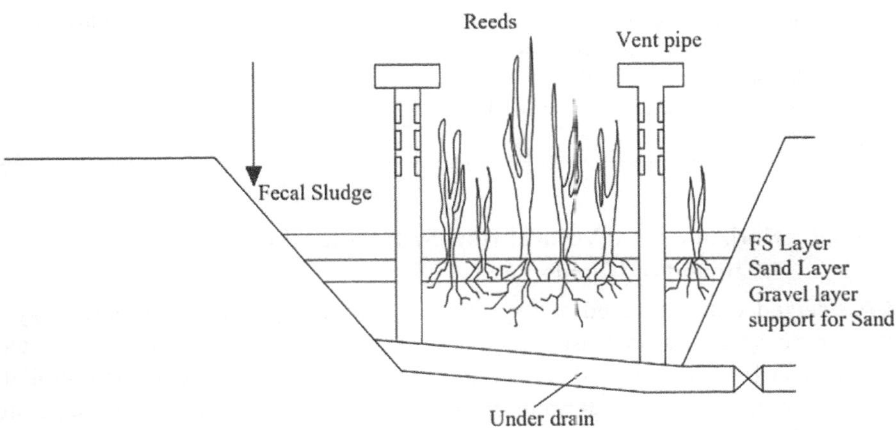

FIGURE 4.6 Vertical flow constructed wetland schematic representation

FIGURE 4.7 Vertical flow constructed wetland. Photo taken by the author

TABLE 4.9
Advantages and Disadvantages of Vertical Flow Constructed Wetland

Advantages	Disadvantages
1. Include dewatering, stabilisation and hygiene requirements in a single treatment stage, unlike all other treatment techniques.	1. Requirement of large area of land
2. Dewatered sludge can be used in agriculture without further treatment or Co composted to yield a high-quality manure/fertiliser	2. Requires care for plant growth.
3. Percolate quality compares favourably to other primary treatments.	3. Care required while removing treated sludge in order not to disturb the media matrix and the vegetation.

4.5.6.2 Horizontal and Vertical Flow Systems for the Treatment of Liquids

Constructed wetlands, reed bed systems, or planted soil filters are natural systems treating low-solid wastewater (Hammer, 1989). This can be pre-treated wastewater from a flush toilet or faecal wastewater from a urine diversion toilet, either combined with wastewater from the kitchen and bathroom, or separate from it (Figure 4.8).

A reed bed is preceded by a settling tank and a distribution tank or a flow regulatory system. It consists of a sand and gravel matrix sealed at the bottom, planted with wetland plants like reeds. Low solid wastewater is discharged from the storage tank on top of the filter or through an underground inlet system and flows through (vertical) the filter. Horizontal flow soil filters are commonly found, and easier to construct than vertical flow filters, but they are less efficient at eliminating nitrogen. Wastewater is treated through several processes, in which bacteria play an important

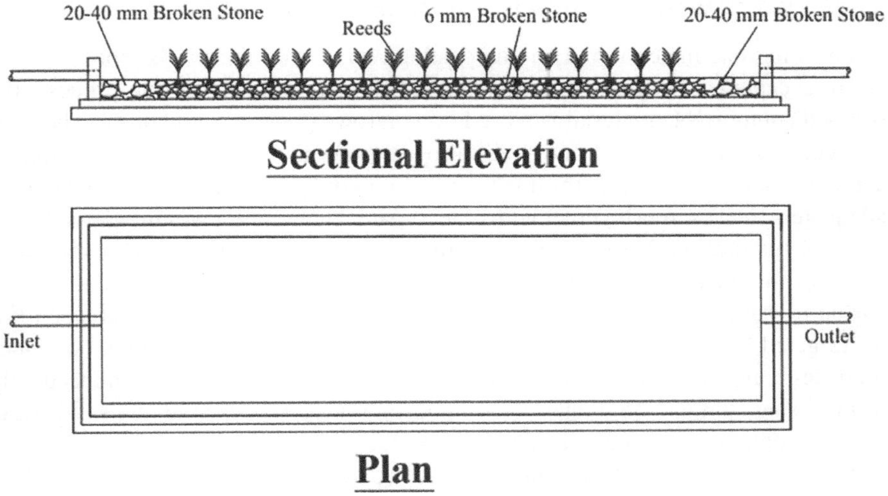

FIGURE 4.8 Horizontal flow constructed wet land schematic representation

TABLE 4.10
Advantages and Disadvantages of Horizontal Flow Constructed Wetlands

Advantages	Disadvantages
1. Easy to operate and maintain.	1. Considerably large area is needed for waste water with high organic load.
2. Removes pathogens from wastewater.	2. Intensive maintenance during first six months.
3. Effluent from wetlands can be used for irrigation.	3. Some degree of post-treatment is needed if the effluent is directly used for edible crop irrigation.
4. Aesthetic appeal.	4. Pretreatment generates some sludge.
5. Operate with low energy consumption.	

role. After treatment, the effluent can be discharged into surface water, used for irrigation, or groundwater recharge.

This system can be used for the following situations.

1. If planted soil filters are applied in hot climates with a continuous growing season, wetland biomass can be harvested.
2. Planted soil filters can be implemented at household or community level. Their use in isolated settlements like rural schools is also possible.
3. Design and construction require a solid understanding of the treatment process.
4. The amount of technical equipment needed is very small.

Costs can vary greatly, and depend, among other factors, on local availability of gravel, the kind of sealing and the cost of land. More details of wetland are given in Chapter 6.

4.5.7 Co-composting with Organic Solid Waste

Composting one material and co-composting two or more materials represent generally accepted procedures to treat excreta. To start the composting process, the blended compostable material is placed in windrows (long or round piles). The 'recipe' combines high-carbon and high-nitrogen materials. Air is added to maintain aerobic conditions, either by turning the windrows or by forcing air through them. To adequately treat excreta together with other organic materials in windrows, the World Health Organisation (WHO), 1989 recommends active windrow co-composting with other organic materials for one month at 328–333 K followed by two to four months curing to stabilise the compost. This achieves an acceptable level of pathogen kill for targeted health values. Adding excreta, especially urine, to household organics produces compost with a higher nutrient value (N-P-K) than compost produced only from kitchen and garden wastes. Co-composting integrates excreta and solid waste management, optimising efficiency (Table 4.11).

Co-composting is applicable in the following situations

1. The type of material, the climate, the amount of space and the equipment and funds available all influence the system design, especially type (e.g. windrow) and size, recipe, and level of technology.
2. Special measures, such as more frequent turning or covering the piles can accommodate extremes of climate or temperature.
3. Composting is a biochemical process, not a bio-mechanical one, and as such requires experience and practical knowledge, together with a high level of management.

TABLE 4.11
Advantages and Disadvantages of Co-composting

Advantages	Disadvantages
1. Flexible approach with highly variable capacity	1. Limited control of vectors and pest attraction.
2. Operation and maintenance require moderate professional experience.	2. Lower cost variants have a high land requirement.
3. Through co-composting, a useful and safe end product is generated that combines nutrients and organic material.	

4.5.8 Anaerobic Digestion

In a digestion process, organic matter from human, animal or vegetable waste is broken down by microbiological activity, in the absence of air. This anaerobic process produces a combustible gas, methane, a source of energy (biogas). The digestion process time may last several weeks to months. The remaining slurry can be removed, either continuously or batch-wise. Several options are available on a household or municipal scale, ranging from simple digestion techniques to technologically complex designs. A domestic anaerobic digestion technique 'fixed dome type' consists of a simple biogas tank with a flat bottom and a round chamber covered with a dome shaped concrete gasholder. The gas is captured in the upper part of the digester. Gas pressure increases with the volume of gas stored, pushing the slurry into a separate outlet tank.

This system is suitable for the following situations:

- Digesters are best suited to warm climates.
- They are most appropriate in rural areas where animal manure can be added to the process.
- The digestion process is sensitive to both temperature and materials. Both need to be controlled.
- Relatively high skills are needed for construction. Operation and maintenance, however, are simple for batch systems.

Capex varies greatly depending on the overall plant concept. The Opex is negligible and if the biogas is put to good use, the Opex will be negative.

TABLE 4.12
Advantages and Disadvantages of Anaerobic Digestion

Advantages	Disadvantages
1. Excreta 'out of sight'.	1. Gas safety risk.
2. Net production of clean renewable biogas.	2. Insufficient pathogen removal without appropriate post-treatment of sludge.
3. Elimination of visual contaminants (e.g. toilet paper).	3. Slurry from digesters has to be removed and treated.
4. Minimal operational control and maintenance.	4. Some control needed.

4.5.9 Biogas Plant

Fresh FS rich in biodegradable organic matter is digested anaerobically, alone or together with animal dung or vegetable waste. The methane produced during the digestion is captured and can be used for cooking, lighting by electricity generation. Liquid effluent and accumulated sludge leaving the digestion process are then treated separately. Design is aimed at biogas production; reactor volumes usually are smaller compared to digesters for sewage sludge. Many different types of biogas reactors have been developed throughout the world. An example is the digester with floating gas dome as it is widely used in India for FS digestion. Following design values are frequently used: solids content (TS) in influent: 5–8%; hydraulic retention time: 30–50 d, organic loading: 1.6–2.2 kg volatile solids (VS) /m^3.day (Figure 4.9)

Digestion of FS for biogas production can be interesting when there is a potential for use of the biogas. Only fresh FS (like from public toilets) is appropriate for biogas production. FS collected from septic tanks, pit latrines, etc., cannot be used directly for biogas production. The sludge needs to have a minimum solids content of 3%. Mixing with animal dung or with vegetable waste may increase solids content and content of digestible organic matter.

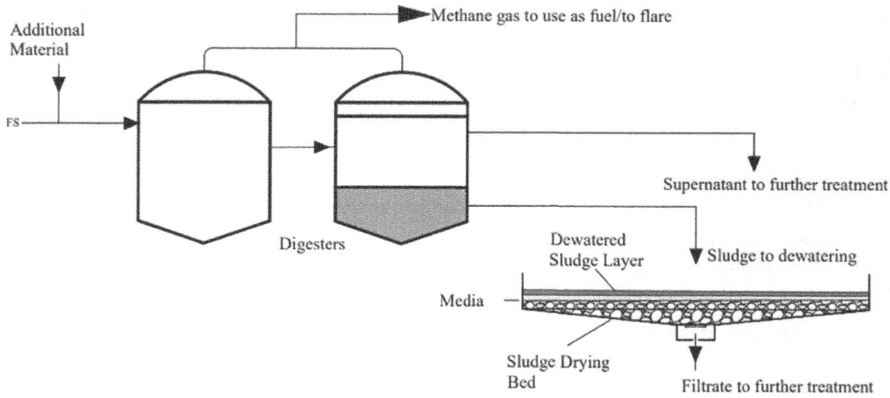

FIGURE 4.9 Biogas plant

TABLE 4.13
Advantages and Disadvantages of Anaerobic Digestion-Biogas Plant

Advantages	Disadvantages
1. Production of combustible gas and generation of revenues.	1. The primary function of a biogas reactor is biogas production rather than sludge treatment. The biogas reactor is therefore generally an additional treatment process.
2. Stabilisation of fresh sludge.	2. Settling in digesters is incomplete and effluents require more extensive treatment efforts than effluents from other primary FS treatment processes.
3. Very less land requirements.	3. The structure is rather expensive, and operation requirements are quite considerable.
	4. Removal of settled and thickened solids can cause difficulties.

4.5.10 Co-Composting and Value Addition

Co-composting and value addition are extensions of the two technologies mentioned above. Constructed wetlands and co-composting. By treating the FS in a VCWL and using the treated FS from the VCWL for co-composting with degradable municipal solid wastes was tried out initially with a year long monitoring of the performance of the treatment plant and quality of the product. This lengthy assessment of the quality of the co-compost and adjustments made during the period yielded a set of results with regard to the ideal ratio of degradable municipal solid wastes to dewatered FS from VCWL. Taking the monitoring a step forward, the growth pattern of various crops to which the co-compost was applied were studied and optimum application rates for best results were found out. In order to meet specific nutrient requirements of certain produces, natural additives like powdered rock phosphate, bio char, etc. were added in prescribed quantities to arrive at optimum quality. The field cases of Ketti Town Panchayat, Adigaratty Town Panchayat, Ponnampatty Town Panchayat, Balangoda Urban Council and Busia Municipality referred to in this book in Chapters 1 and 3 were the success stories of this approach. There are other places

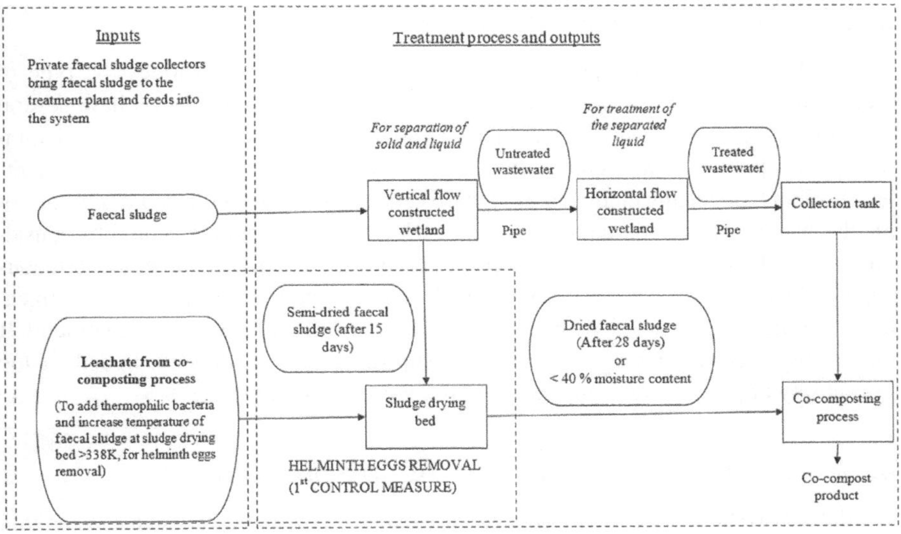

FIGURE 4.10 Process flowsheet for co-composting

where the same approach is being replicated which include various towns in Sri Lanka, India and East Africa.

4.6 SELECTION OF TECHNOLOGY AND DECISION-MAKING PROCESS

4.6.1 Characteristics of the Faecal Sludge to Be Treated

The characteristics of the faecal sludge (FS) varies widely according to circumstances and sources. In some cases, fully digested material would be the major component. But, the effluent in other cases, the undigested material may contribute significantly to the contents. In the case of former, methods adopted would be focused on dewatering while in the latter, the approach to treatment would be like that of sewage. For fully digested sludge or FS with a major component of fully digested sludge, the technologies considered are limited. However, the potential for co-composting partially or fully treated FS with degradable municipal solid wastes (MSW) adds another layer to selection of technology.

4.6.2 Flow Pattern and Flow Management Options

Normally, flow management in FSM is a non-issue since the treatment of fully digested sludge would invariably be a batch process irrespective of whether anaerobic digestion, sludge drying bed or constructed wetland is the technology selected. If a treatment system resembling a Sewage Treatment Plant (STP) is adopted to the presence of high quantities of fresh and partially digested faeces, the flow management options described in sewage treatment with a combined stream of influents would be applicable for FS too.

4.6.3 Availability of Power/Source of Power

Availability of power and proximity to a source of power determines the type of technology to be adopted. All aerobic processes and electrocoagulation have several equipment as part of the system which need a source of power in the vicinity and uninterrupted supply of power. Quite often, a generator would be provided in such plants as a backup to meet any emergency situation caused by a power outage. Possibilities of an alternate perennial source of green energy like solar or wind too should be considered at the planning stage of the project. In the absence of a good source of continuous supply of electricity, the process to be opted would be multi stage anaerobic treatment and sometimes constructed wetlands subject to availability of adequate area of land. If dewatering or dewatering coupled with co-composting is the method adopted, there too availability of power is relevant although to a smaller degree.

4.6.4 Land Requirements and Financial Considerations

Availability of land has a direct impact on selection of technology. It can be seen that ponds, a completely low maintenance system needing no operators either requires very large tracts of land due to the high retention time required. Stabilisation ponds,

TABLE 4.14
Land Requirement for Partially Digested + Raw Sewage Using Low Maintenance Systems (Low Sludge Generation-Only Periodical Removal of Sludge)

S. No	Description	Area Required ($m^2/m^3/d$)	Nature of Treatment	Remarks
1	Septic tank	0.5–1.0	Primary, anaerobic	Shall need more treatment including a secondary before disposal of both effluents and sludge.
2	UAF	0.5–1.0	Secondary, anaerobic	Shall need more treatment before disposal of both effluents and sludge.
3	ABR/ISBR	1.0–1.5	Primary, anaerobic	Shall need more treatment before disposal of both effluents and sludge.
4	VCWL	10–15	Natural, primary	Needs only physical treatment before the process. The sludge can be processed further – either dewatering or co-composting. Effluent may not need treat further treatment before discharge.
5	HCWL	5–30	Natural, primary/ secondary after an anaerobic treatment or a VCWL	Needs physical treatment and preferably a primary anaerobic treatment before the process. Possibility of sludge formation is quite remote. Effluent may not need further treatment before discharge/reuse.

trendy in the past, are no longer are seen as a serious option for municipal sewage treatment due to the large area of land to be committed and the potential for public nuisance like mosquito menace and generation of malodourous gases. If sufficient land is not available, such a large system cannot be considered. Although much smaller in size in comparison to ponds, constructed wetlands too need large tracts of land in comparison with biochemical reactors like Anaerobic or Aerobic Reactors. Within the group of aerobic technologies, the attached growth processes have smaller footprint. Technologies like membrane bio reactor (MBR) and electrocoagulation (EC) have still smaller footprints occupying a very small area in comparison. It may be of interest to note that smaller the foot print of the technology adopted, higher the Opex would be. For example, electrocoagulation needs the smallest of area in comparison to any other technology but has a very high Opex both in terms of power charges and cost of frequent replacement of the sacrificial electrodes. Further, in order to achieve a certain amount of economic viability, Co-composting with degradable MSW would become part of the system since dewatered FS hardly has sale value as manure. The land requirements will vary proportionately and hence it is easy to arrive a general figure. However, the capital investment (Capex) and operation and maintenance (O&M) expenses vary widely and are not proportionate to the capacity. Hence a certain arbitrary capacity and characteristics are assumed for indicating the kind of investment and financial outlay is needed.

For FSTPs to treat raw or mixtures of raw excreta and partially digested sludge, the technology used is similar to sewage treatment plants. Quite often FS is found with very high COD in which case a series of anaerobic reactors with or without gas recovery would be needed at the first stage of biochemical reactions. The preliminary and primary treatment consisting of physical operations for removal of large solids, grit, grease, etc., are more or less the same in all methods of treatment. The critical difference is the foot print of the biological reactor.

The tables of comparison of land area needed and operation and maintenance costs (Opex) for low maintenance technologies are given in the Tables 4.15–4.17. As is well-known, the Capex does not have a proportional relation to the feed flow rates and tends to taper as the capacity goes up. The Opex is generally almost proportionate to the capacities except in the case of human resources deployed. These

TABLE 4.15

Opex for Partially Digested + Raw Sewage (Low Maintenance Systems Low Sludge Generation – Only Periodical Removal of Sludge) Basis 1 m³/d

S. No	Description	Opex	Remarks
1	Septic tank	Negligible	Cost of emptying only is required. Frequency of emptying would be once in two to five years.
2	UAF	Negligible	Cost of emptying only is required. Frequency of emptying would be once in two to five years.
3	ABR/ ISBR	Negligible	Cost of emptying only is required. Frequency of emptying would be once in two to five years.
4	VCWL	Negligible	Periodic trimming of reeds, replanting if necessary and removal of dried sludge need unskilled operators.
5	HCWL	Negligible	Periodic trimming of reeds and replanting if necessary.

are indicative figures and should not be taken as absolute figures. The reasons for variations in Capex can include but are not limited to

- Site conditions like
  - Lay of the land like flat, sloped terraced, etc. (for multi-stage treatment, sloped land would facilitate gravity flow while a flat land would need additional constructions/earthwork excavation/pumping systems, etc.).
  - Soil characteristics which in turn influence materials of construction.

TABLE 4.16
Land Area Requirement for Fully Digested Sludge

S. No	Description	Area Required ($m^2/m^3/d$)	Nature of Treatment	Remarks
1	VCWL	6–18	Natural treatment + dewatering	Screening of feed required. Treated sludge fit for disposal/reuse/co-composting. Frequency of cleaning flexible. The effluent may need further treatment before disposal or reuse.
2	Conventional sludge drying bed	2–5	Dewatering	No pre-screening required. Needs cleaning when dry. Treated sludge fit for disposal/reuse. The effluent shall need further treatment before disposal or reuse.
3	Plate and frame filter press	Negligible	Mechanised dewatering	Dried sludge fir for reuse or disposal. Filtrate shall need more treatment before reuse or disposal. The filter clothes need cleaning and drying before reuse.

TABLE 4.17
Opex for Fully digested Sludge Basis 1 m^3/d

S. No	Description	Opex	Remarks
1	VCWL	Negligible (only unskilled operators)	Screening of feed required. Treated sludge fit for disposal/reuse/co-composting. Frequency of cleaning flexible. The effluent may need more treatment before disposal or reuse.
2	Conventional sludge drying bed	Negligible (only unskilled operators)	No pre-screening required. Needs cleaning when dry. Treated sludge fit for disposal/reuse. The effluent shall need more treatment before disposal or reuse.
3	Plate and frame filter press	High (skilled or semi-skilled operators needed)	Power, Maintenance cost of pump, Periodic cleaning and occasional replacement of Filter clothes, Needs a minimum of 1 skilled operator and 1 unskilled operator.

    - Availability of construction materials.
    - Access and ease of transport of materials.
    - Availability of skilled labourers.
- Characteristics of the feed and the stages required based on the characteristics of the feed.
- Degree of treatment required.
- End use of the final effluents.
- End use of the sludge generated from the process and the type of further processing required.

TABLE 4.18
Area Requirement for Partially Digested + Raw Sewage Mechanised Systems (High Sludge Generation Needs Regular Removal of Sludge)

S. No	Description	Area Required ($m^2/m^3/d$)	Nature of Treatment	Remarks
1	ASP	0.4–0.5	Secondary, aerobic	Needs Physical Primary Treatment and post reaction sedimentation, disinfection and filtration. Effluent fit for discharge/reuse. Sludge can be dewatered and disposed/reused.
2	SBR	0.25	Secondary, aerobic	Needs physical primary treatment and post reaction disinfection and filtration. Effluent fit for discharge/reuse. Sludge can be dewatered and disposed/reused.
3	MBBR	0.1–0.15	Secondary, aerobic	Needs physical primary treatment and post reaction sedimentation, disinfection and filtration. Effluent fit for discharge/reuse. Sludge can be dewatered and disposed/reused.
4	SPBR/SAFF	0.15–0.25	Secondary, aerobic	Needs physical primary treatment and post reaction sedimentation, disinfection and filtration. Effluent fit for discharge/reuse. Sludge can be dewatered and disposed/reused.
5	MBR	0.08–0.12	Secondary, aerobic	Needs physical primary treatment and post reaction disinfection. De-chlorination may be needed depending on the means of disposal and reuse. Effluent fit for discharge/reuse. Sludge can be dewatered and disposed/reused.
6	EC	0.06–0.1	Electrolysis	Needs physical primary treatment and post reaction filtration. Effluent fit for discharge/reuse. Sludge is negligible which can be reused/disposed.

TABLE 4.19
Opex for Partially Digested + Raw Sewage Mechanised Systems (High Sludge Generation Needs Regular Removal of Sludge) Basis 1 m³/d

S. No	Description	Opex
1	ASP	Only routine Opex involving power, spares, and chemicals. No media for replacement.
2	SBR	Only routine Opex involving power, spares, and chemicals. No media for replacement.
3	MBBR	Only routine Opex involving power, spares, and chemicals. Media may not need replacement for decades.
4	SPBR/SAFF	Only routine Opex involving power, spares, and chemicals. Media may not need replacement for decades. Occasional cleaning of media may be required.
5	MBR	Routine Opex low. High cost for periodical replacement of membrane.
6	EC	Low footprint and number of equipment and easy to operate. Cost of replacement of the sacrificial anodes would be high.

- Possibility of forward integration and/or value addition.
- Access to the point of disposal/reuse/value addition.

In view of the above the figures in the tables are to be used quite judiciously and in a realistic manner.

A set of indicative figures and information on the area requirement, Capex and Opex for treatment of the two different types of faecal sludge and different types of treatment systems are given below. These tables can be used as an indicative guide for decision makers while planning a faecal sludge management system for a small town. (The technologies like ASP, MBR, MBBR, UAF, and EC listed in the Tables 4.14–4.19 are explained in Chapter 5.)

Irrespective of the technology adopted, FSM requires semi-skilled and skilled operators. High level of safety measures too is needed to be ensured in the case of FSM. The operators needed would be either skilled or at least semi-skilled. Even if the technology chosen is a low maintenance system, the importance of the quality of operators cannot be overemphasised. The Opex has to be budgeted suitably to include the cost of employing not only skilled or semi-skilled operators but supervisory staff with credentials too.

When a technology that involves extensive Operation & Maintenance protocols, that involves not only a cost implication but a management system in place to ensure smooth functioning of the plant. The factors involved include budgeting for recurring costs for manpower, maintenance of the equipment, recruitment policy for the personnel, supervisory functions and allocation of such functions to someone in authority and ownership of the plant, etc.

4.6.5 Climate Impact and Sustainable System for Monitoring

While no treatment has a higher impact on climate than in the case any treatment, the technology with the least implications on climate among the various options need to be selected. However, factors mentioned above too will have to be considered in tandem with climate implications. For example, in cases where land is available in plenty, when there is a choice between mechanised systems and constructed wetlands, it can be seen that a constructed wetland is a better choice from the climate angle since the methane generation in a CWL is negligible or even nil. The vegetation in the CWL almost offsets any carbon dioxide generated there. A CWL is almost climate neutral a technology. The low maintenance option of anaerobic process releases methane while the aerobic processes release only Carbon dioxide which has a lower climate impact than methane. The ecological footprint of a CWL is much lower than any other technology in terms of use of finite resources. While mechanised technologies use power and equipment with moving parts, anaerobic processes have a significant climate impact in terms of the greenhouse gases. It can be seen that CWL is the greenest of technologies available.

Since the final product from FSM is either soil conditioner or co-compost or fortified manure, strict monitoring protocols need to be in place. FSM has a good potential to ensure a continuous stream of revenue income, that comes with a price tag of cost of monitoring. The monitoring would entail safety, health and environment (SHE) conditions at the project site and quality assurance of the final product. Since there is a marketing angle involved, customer feed backs and follow up too would become part of the activity.

4.7 WORLD HEALTH ORGANISATION (WHO) GUIDELINES

World Health Organisation (WHO, 2006) has issued global guidelines and tools for wastewater reuse in agriculture. The new guidelines take due recognition of the fact that effective treatment of wastewater may not materialise in some areas for many years. The draft guidelines are pragmatic and the approach is based on scientific consensus and best available evidence. They are meant to be adapted to local social, economic, and environmental conditions. The approach is based on risk/benefit analysis developed around good practices and a multiple barrier approach. The guidelines strive to maximise overall public health sector benefits and at the same time beneficial use of scarce resources. Demand for waste water as a reliable source of water and nutrients for agriculture is on the rise due to population growth, urbanisation, increasing water scarcity, and the effects of climate change.

The available measures for health protection can be grouped into five main categories:

1. Waste treatment
2. Crop restriction
3. Irrigation/application technique
4. Human exposure control
5. Chemotherapy and vaccination

It will often be desirable to apply a combination of several methods. Sometimes partial treatment to a less demanding standard may be sufficient if combined with other measures such as crop restriction but it may need to be supplemented by additional measures to protect agricultural workers. An approach towards acceptable local risk limits, applicable for sanitation systems where the use of the excreta products is practiced, is related to subsequent reduction or increase in the prevalence of infections. The goal will be to reduce the number of infections by implementing sanitation *per se* including introducing new, more efficient treatment or exposure reduction alternatives, combined with other interventions related to safe treatment and storage, hygiene/health education as well as provision of safe water supply. The following guidelines are to be adhered to for the safe use of wastewater when used as fertiliser for agriculture. Excreta and faecal sludge should be treated before it is used as fertiliser and the treatment methods validated (Prescod, 1992).

1. Treated excreta and faecal sludge should be worked into the soil as soon as possible and not be left on the soil surface.
2. Improperly sanitised excreta or faecal sludge should not be used for vegetables, fruits, or root crops that will be consumed raw, excluding fruit trees.
3. A withholding period between fertilising and harvest apply for treated excreta and faecal sludge. This period should be at least a month.

TABLE 4.20
Degree of Health Protection in Relation to Application of Faecal Sludge

Category A – Protection Needed Only for Field Workers	Category B – Further Measures May Be Needed	Category C – Treatment to 'Unrestricted' Guidelines Is Essential
Crops not for human consumption (e.g. sisal and cotton)	Pasture, green fodder crops	Crops often eaten uncooked and grown in close contact with wastewater effluent (e.g. lettuces or carrots, or spray-irrigated fruit)
Crops normally processed by heat or drying before human consumption (grains, oilseeds, and sugar-beet)	Crops for human consumption that do not come into direct contact with wastewater, on condition that none must be picked off the ground and that spray irrigation must not be used (tree crops, vineyards, etc.)	Landscape irrigation with public access (parks and lawns)
Vegetables and fruit grown exclusively for canning or other processing that effectively destroys pathogens	Crops for human consumption normally eaten only after cooking (potatoes, eggplant, and beetroot)	
Fodder crops sun-dried and harvested before consumption by animals	Crops for human consumption, the peel of which is not eaten (coconut, melons, citrus fruits, bananas, nuts, and groundnuts)	
Landscape irrigation in fenced areas without public access (nurseries, forests, and green belts)		

Source: WHO (2006).

TABLE 4.21
Degree of Treatment Required for Applying Methods for Irrigation

High-Medium Treatment Required	Low Treatment Required
Border (flooding) irrigation	Subsurface irrigation
Furrow irrigation	Localised irrigation
Sprinkler irrigation	

TABLE 4.22
Human Exposure Control per Group

Group Exposed	Control of Human Exposure
Agricultural field workers and their families	Protective clothing, personal hygiene, sanitation, immunisation, chemotherapy
Crop handlers	Protective clothing, personal hygiene, sanitation, immunisation
Consumers (of crops, meat, and milk)	Cleaning vegetables, cooking foods thoroughly, high standards of personal and kitchen hygiene, inspection of foods, halting waste application two weeks before harvest
Those living near affected fields	Signs, fences, no spray irrigation near residences (50–100 m)

In places which have water scarcity wastewater is treated and used in industries as cooling water, boiler feed water, process water, and for heavy construction. Reclaimed water is ideal for many industrial purposes where water of high quality is not necessary. Industries situated near to treatment facilities of populated areas are already reusing the reclaimed water. The concerns are scaling, corrosion, biological growth, and fouling. Appropriate treatment options are to be adopted to overcome the problems associated with it.

5 Sewage Treatment

5.1 INTRODUCTION

In most of the small towns in developing countries, there may not be any sewer system and a centralised sewage treatment plant. In such cases, residences, commercial establishments, and public offices would be adopting in situ sanitation systems. In situ sanitation would imply use of a septic tank or leach pit or similar alternate systems for the substructure which get emptied periodically. The contents of such substructures could be either fully digested or partially digested and may even contain fresh faecal matter. The terms sewage and wastewater are used interchangeably in the case of municipalities since the wastewater stream generally contains both sewage and sullage. In small communities, it is possible to separate the two streams and provide separate initial treatments before combining them for the subsequent stages of treatment.

There could be situations in which a small cluster of the town is sewered and connected to a sewage treatment plant and the rest of the area having in situ sanitation facilities. In such cases, too, the description of the various technologies in this chapter is of relevance since there would be places of public congregation like apartment complexes, shopping malls, hospitals, educational institutions, places of worship, conference halls, museums, and government offices where in situ sanitation systems would be inadequate due to the large volumes of sewage and sullage to be treated and disposed or reused. There are statutory requirements in most of the countries with respect to the degree of treatment, means of disposal, and standards for reuse of the treated sewage. While these standards may vary from country to country, there would be standards developed and enforced by the statutory bodies like Pollution Control Boards or Department of Health or Environment in that country.

The method of treating raw sewage will have the following steps:

- physical preliminary treatment
- physical and if necessary chemical primary treatment
- secondary treatment which generally is biological treatment followed by settling
- tertiary treatment of disinfection and filtration of the treated effluent
- dewatering of the sludge for reuse or disposal

The objective of the treatment plant is to achieve prevailing legal standards of water quality for discharge or reuse. There is a long list of liquid waste treatment methods available (Metcalf and Eddy, 2014) which can be combined to have an appropriate system to achieve the required level of contaminant removal. The sullage component may receive fugitive effluents from agricultural fields

DOI: 10.1201/9781003650737-5

or micro-level industries in which case, more rigorous treatment for removal of specific chemical contaminants would be required. The sludge resulting from sewage treatment operations is treated by various methods in order to reduce its water and organic content and make it suitable for final disposal and reuse. Typically, the following are among the popular technologies adopted in various stages of treatment (Table 5.1).

Additionally, there are methods like electrocoagulation, constructed wetlands, and nutrient film technology. There are some not so popular technologies like Imhoff tank which are excluded from this list. Some combinations of these methods have become popular, such as decentralized wastewater treatment systems (DEWATS) and low maintenance water treatment systems (LOMWATS), but these are not new or original 'treatment methods', but just combinations of known treatment methods and hence basically open-source technologies. This section would expand on low maintenance technologies since the focus of the book is on small towns in developing countries.

TABLE 5.1
Classification of Sewage Treatment Methods

Types of Methods	Process	
Physical methods	Screening	
	Comminution	
	Sedimentation	
	Flotation	
	Granular-medium filtration	
Chemical methods	Flocculation	
	Coagulation	
	Chemical precipitation	
	Chlorination	
	De-chlorination	
Biological methods	Anaerobic processes	Septic tanks
		Up-flow anaerobic filters
		Anaerobic baffled filters (ABR) and variants
		Up-flow anaerobic sludge blanket reactor (UASB)
		Anaerobic ponds
		Biogas
	Aerobic processes	Aerated lagoons
		Activated sludge process and variants
		Trickling filters
		Rotating biological contactors
		Submerged packed bed reactors/submerged fixed film reactors
		Moving bed bioreactors
		Membrane bioreactor
		Stabilisation ponds

5.2 PRETREATMENT UNITS

The methods used are screening, sedimentation, or floatation. Screening is the process by which particles are physically strained out. Heavier particles with a density greater than that of water are settled out using gravity. These settled particles will form a layer at the bottom of the vessel. This settled solid volume is called sludge. Naturally, particles settle when there is no turbulence and the volume is left undisturbed for some time before cleaning the settling tank. The frequency of cleaning is dependent on the rate of deposition of the solids. The same principle is utilised in all sedimentation tanks. Addition of various chemicals for size agglomeration can facilitate faster sedimentation. Particles lighter than water are made to float on the surface and skimmed off. Aeration can facilitate faster flotation since the air bubble decreases the density of the material to be removed.

5.2.1 SCREEN CHAMBER

Screen Chambers are provided at the inlet to a treatment plant. The purpose of screening is to remove larger, inorganic debris from the waste stream.

Screens are provided to

- prevent clogging of pipelines
- protect downstream mechanical parts from damage
- ensure the efficient operation of succeeding units
- eliminate unsightly floating matter.

In larger plants, bar screens are made by arranging bars in slanting position with spacing in between them. The larger particles will get stuck in-between the bars and water will flow through space between bars. The screened materials are usually disposed in sanitary landfills or by other processes in the Solid Waste Management Centre. For easy cleaning, the screen buckets or trap like screens are used in smaller sewage treatment plants. The screens are classified as coarse, medium, and fine screens based on the spacing between the strainers. The normal classification is as follows.

- Fine screens – spacing under 10 mm
- Medium screen – spacing of 10–40 mm
- Coarse screen – spacing of over 40 mm

Screens are given specifications based on the opening size. A 60 mm screen means the spacing between the bars is 60 mm. The material of construction of the screens is important due to possibilities of corrosion over a period of time. Bar screens are generally made of stainless steel or mild steel with corrosion proof coating. Other possible materials of fabrication are aluminium and PVC. Corrosion control of MS bars is done with corrosion proof paints, rubber coating, or epoxy coating after preparing the surface by abrasive blasting, generally sandblasting.

FIGURE 5.1 Bar screen for removal of large floating materials

Wastewater will enter the screen chamber through an inlet pipe with a particular flow velocity. Floatable solids which are larger than screen openings will get strained and build up ahead of the screen. The flow will reduce as lesser flow area is available due to the buildup of strained-out materials before the screen. This will enhance the muck and sediments to settle down. This will again act as a screening bed and facilitate more screening and sedimentation. The screen and the sediments in upstream have to be cleaned periodically to facilitate the proper functioning of the screen chamber. This process is called raking. Platform with weep holes is provided for the operator to perform the raking operation. This platform is also utilised as a drip tray for the collected debris. Regular cleaning is required for proper screening operations. In larger plants, there are mechanically operated raking devices which are placed over the platform over the screen. There will be a level sensor and the raking mechanism will become functional when the water level reaches a predetermined level. The wastewater free of any floatable material will flow to the next units.

5.2.2 Grit Chamber cum Grease Trap

Grit is the inorganic particles of sand, gravel, cinders, and pebbles which don't get removed by screens. The grit removals are performed to

- Protect moving equipment from abrasion
- Reduce deposition in pipelines and channels
- Reduce frequency of digester cleaning

Oil and grease tend to have acidic pH and have bactericidal/bacteriostatic effects too which would affect the chemical and biological processes downstream. Oil and grease can interfere with working of mechanical units too. Hence, a need to remove them before the effluent is sent to the next unit of treatment. The oil and grease separated can be skimmed off from the surface of the tank. Grit Chamber is a type of sedimentation tank designed to settle out heavy grit particle by gravity. Lighter oil and grease particle are made to float and removed from the surface of skimming tanks. In small plants, these two processes will be combined and performed in a single unit. These chambers are designed providing more depth and low surface area. The velocity of flow to the unit will be controlled by the design. Due to the sudden fall of wastewater to the trap like arrangement the oil and grease will get separated. These will form a surface layer. The chamber is designed deep enough to settle the heavy grit particles without any hindrance. The grit, denser than water will settle down. These can be collected from the bottom of the unit. A baffle arrangement will keep away the oil or settled grit from escaping to the outlet. Sometimes lamella plates are arranged to increase the efficiency of the settling of grit. Skimming has to be done periodically since the fat and grease layer may cause multiple problems. In small plants, this tank is cleaned manually while larger plants adopt mechanical skimming.

5.2.3 Dissolved Air Floatation (DAF)

This is an improved grit and oil separation unit. Dissolved air under pressure is supplied to enhance the removal process. In this process, both grit and oil are made to float and are together removed by skimming arrangements. Dissolved air flotation (DAF) involves the dissolution of air in wastewater by pressurising it in a pressure vessel. When the air-saturated wastewater is released to the flotation tank, the sudden decrease in pressure causes the air to come out of solution as microbubbles

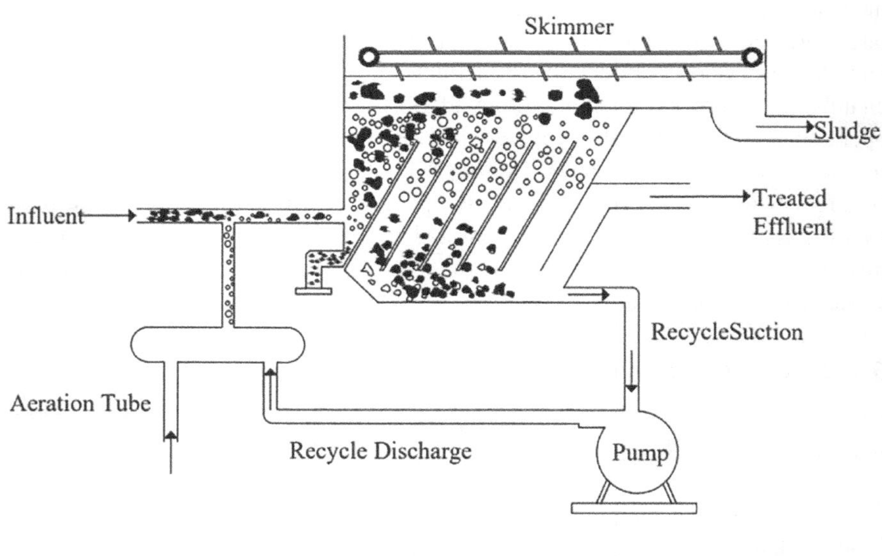

FIGURE 5.2 Schematic diagram of dissolved air floatation unit for removal of oil and grit

similar to fizzing action in carbonated drink bottles. The microbubbles produced will attach themselves to solid and oil particles in the wastewater and make them float. The result is a floating mat of sludge on the surface which is skimmed off. If sewage and sullage are received as separate streams, the DAF will be used in the sullage line which would be receiving oily matter from bathrooms, kitchen, etc. The use of DAF is particularly useful in the case of ayurvedic treatment centres where the use of mixtures of different types of oils is common. Use of DAF drives out objectionable gases also. The addition of chemicals like alum and polyelectrolytes are practiced quite often in DAF for improving the efficiency of the system (Figure 5.2).

5.3 PRIMARY TREATMENT

After removing the solid matter, grit, oil, and grease by preliminary treatment, the next step is removal of suspended solids. This is done in primary treatment and mainly by sedimentation methods as it is most practical and economic process.

5.3.1 Primary Settling Tank (PST)

Settling tanks are provided before biological treatment, hence the name primary settling tank (PST). Sedimentation will help to remove Total Suspended Solids (TSS) by around 50–70%. If separate streams of sullage is available, facilities for chemical treatment and flocculation would be provided. In the case of a combined stream of raw effluents, chemical treatment is not done normally. The primary settling tanks help protecting the subsequent units from possible mechanical wear by settleable solids and reduce the BOD load by 20–40%. The general principle of all sedimentation tanks is the same. Avoiding turbulence and providing enough settling time is all that is needed for a settling tank to function properly.

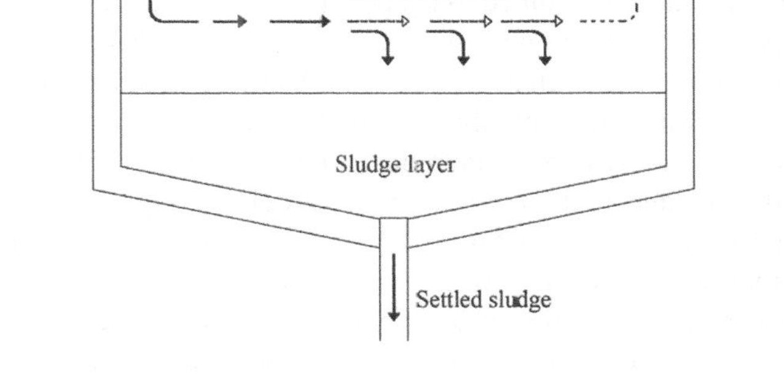

FIGURE 5.3 Primary settling tank

Lighter particles would float upward and form a layer at the top of settling tank which need to be skimmed of regularly. The concentrated solid portion settled at the bottom is sludge. Sedimentation tank bottom will be usually hopper shaped (conical) to facilitate the efficient collection and removal of the settled solids. Sludge pumps or gravity flow are used for desludging.

The sedimentation tank has three-zone

- Inlet zone: With arrangements to distribute the flow without turbulence.
- Sedimentation zone: Where settling happens and the settleable solids get deposited as sludge. Sludge collection arrangements are to be provided here.
- Outlet zone: Portion where the effluent is drawn out of the sedimentation tank. Arrangements should be provided to collect the effluent without scum and sludge.

In long rectangular tanks, inflow is introduced at one end and overflow is collected at the other end or at a different point along the length. An endless conveyor scrapes the floating material. Inlet zone will be provided with baffle arrangements to distribute the influent without creating turbulence. The sludge getting deposited in the middle settling zone will be scraped into a collection trough provided at the inlet of the tank. Sludge pumps will suck out the collected sludge and deliver it to solid waste processing units. The efficiency of sedimentation can be enhanced by providing inclined plates (lamellae) and tube.

5.3.2 Clariflocculator

Flocculation is a chemical process where coagulants are added to enhance floc formation and thus aid sedimentation process. The coagulants are surface-active agents which will result in the formation of large settleable mass. These chemicals are initially mixed with the wastewater to be treated in a flash mixer or a Parshall flume. Heavier the particles, easier the sedimentation will be. Mixing has to be provided for proper floc formation. Then these chemical flocs are sedimented out by gravity in a sedimentation tank called clarifier. Clarification is the process of removing this chemical floc. These processes are carried out combined in a unit called clariflocculator. This unit will help to remove out the good portion of nutrient content phosphate. In smaller plants, separate units for flocculation and clarification won't be found. The flocculation is assumed to happen in the chemical mixing stream and the unit will consist of a clarifier unit. The sludge so formed is chemical sludge which is to be disposed through specialist agencies or to Sanitary Landfills. The processes of flocculation and settling by combining the flocculator and the clarifier are done in the clariflocculator. It is more popular in municipal water treatment plants and industrial effluent treatment plants than in sewage treatment plants (Figure 5.4).

Wastewater dosed with coagulant will enter the tank through the inlet provided at the top, mounted over a supporting column at the centre of the tank. The inlet is provided with energy dissipating arrangements and a barrier like arrangement called feed-well to prevent turbulence due to the flow. These arrangements will create an undisturbed condition for the sedimentation process to take place. The floor will be

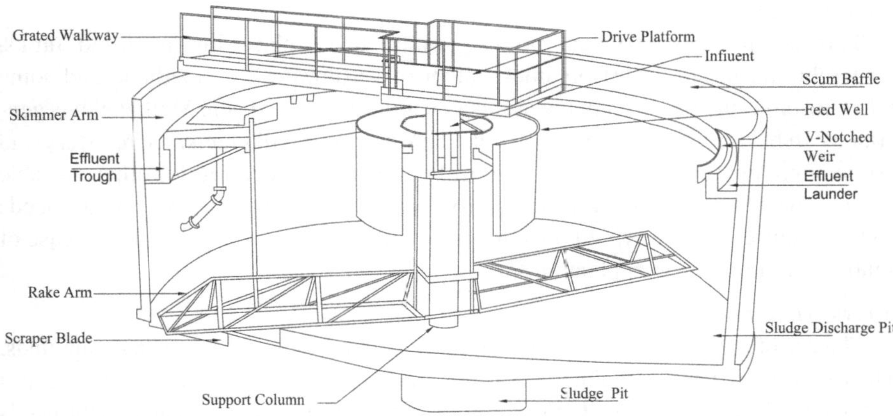

FIGURE 5.4 Clariflocculator which combines the operations of flocculator and clarifier

sloping towards the centre and will be provided with two blade-like wings called scrapper mechanism. The slow movements of these blades will help in the floc formation. Using some baffles, the tank is designed to keep out the scum and collect the clear supernatant by V-notched weir overflow from the top portion of the tank. The overflow rate is fixed ensuring proper impoundment time for successful sedimentation process to happen. There will be a skimmer arm rotating at the water level and this will skim and dispose of the oil layer to the scum trough. The bottom revolving scraper blades have wipers attached which will also help in accumulating and collecting the chemical sludge on the tank floor. The sludge will be collected in a sludge pit from where it can be taken for further treatment.

5.3.3 Equalisation Tank

Equalisation Tank does not offer any treatment per se and essentially is a flow management system which holds the inflow to achieve flow equalisation, mixing, and homogenising. Continuous uniform flow to the biological reactors can be ensured by providing adequate holding capacity for the Equalisation Tank. Equalisation Tanks are generally provided with a nominal aeration facility for mixing and homogenisation as well as a means of odour control. There are two types of aeration used in practice. One is to provide the tank with floating surface aerators. These move up and down the tank along with the level of the liquid in it. In STPs provided with air blowers, the same air blower would be used as a source of compressed air and diffused into the equalisation tank through a set of coarse bubble diffusers. In some case, providing mechanised agitators too is practiced.

5.4 SECONDARY TREATMENT

The major treatment in a sewage treatment plant is the reduction of BOD and COD load of wastewater. The process adopted is biological except in the case of electrocoagulation.

Mainly, there are three types of biological treatment methods.

Anaerobic Treatment

This is an energy-efficient process which takes place inside closed tanks. Anaerobic bacteria convert organic content of wastewater into methane and some other unstable end products in the absence of atmospheric oxygen. Methane thus generated can be used if it is in large quantity. However, in small plants, the recovery and use may not be economically viable. Post treatments are required for other unstable end products. This treatment process proceeds at a slow rate but low energy needs and no requirement for skilled operators make this an attractive option in the case of small community-level applications.

Anoxic Treatment

The use of anoxic reactor is primarily intended for removal of nitrogen compounds. The dissolved oxygen is maintained at less than 1 mg/L in this reactor. The nitrates in the raw effluent get converted in to nitrogen gas in this process. The reactor needs a mixing facility to maintain uniformity. This is achieved by recirculation in small plants and an agitation system in larger plants. Denitrifying bacteria are sensitive to temperature and pH. Optimal temperatures for a proper nitrification–denitrification reaction are typically in the range between 288 K and 298 K. The ideal pH for best results is around 7. The denitrification process consumes alkalinity. Hence, maintaining adequate levels of alkalinity in the water is essential to prevent pH drops that could harm bacterial activity. Deviations from these conditions can slow down or inhibit the denitrification process. Denitrification requires a significant amount of organic carbon. This can come from the wastewater itself under normal circumstances. In exceptional cases, the carbon requirement needs to be supplemented externally. Common carbon sources include methanol, acetate, and other readily biodegradable organic substrates. The hydraulic retention time (HRT) must be carefully controlled to provide sufficient time for denitrification while avoiding excessive tank volumes.

Aerobic Treatment

Aerobic microorganisms convert organics into carbon dioxide and new biomass in the presence of oxygen. The end products of this process are stable. Aerobic microorganisms require oxygen and so air must be continuously circulated through the tanks. This treatment requires large-volume tanks and various mechanical equipment to facilitate aerobic bacterial functions. While the capital investment and operation costs are more than in the case of anaerobic processes, the footprint of the plant would be smaller. The sludge production in aerobic processes would be several times that of anaerobic processes but handling of aerobic sludge is easier than anaerobic sludge. Quite often an anoxic reactor is used in tandem with aerobic reactors which aims to reduce nitrogenous content by a nitrification–denitrification process in a low-oxygen environment.

Both biological systems can be again classified based on the way the bacterial masses are grown inside the reactor.

- Suspended growth process: In these reactors, microbial flocs are kept in suspension.
- Attached growth process: In these reactors, bacteria are grown adhering to an inert media.

5.4.1 Anaerobic Process

The anaerobic process takes place in the absence of dissolved oxygen. Anaerobic treatment is preferred for wastewater containing high COD. The anaerobic process is a multi-step process. A diverse group of microorganisms degrade the organic matter in a sequential order.

- **Hydrolysis**: In this stage, long-chain chemical compounds are broken down to simpler compounds. Carbohydrates will be converted to soluble sugar units and proteins to amino acids.
- **Acidogenesis:** In the second stage, microorganisms break down the single molecules of sugar and amino acids even further into ethanol and fatty acids, as well as producing carbon dioxide and hydrogen sulphide as by-products.
- **Acetogenesis:** In the third stage, the ethanol and fatty acids are converted into hydrogen, carbon dioxide, and acetic acid.
- **Methanogenesis:** In the fourth and final stage, microorganisms convert the remaining hydrogen and acetic acid into methane and more carbon dioxide.

The main types of anaerobic bacteria involved in these anaerobic units are acid-forming bacteria and methane-forming bacteria. The survival conditions and by-products of these bacteria vary depending on the reactor condition and characteristics of the feed.

The essential conditions for efficient anaerobic treatment

- Devoid of excessive air/Oxygen
- COD should be in the range of 350–1000 mg/L.
- Presence of enough nutrients (Nitrogen and Phosphorus) and trace metals
- No toxic/inhibitory compounds present in the influent.
- The pH range is between 6.8 and 7.2.
- Sufficient alkalinity should be present.
- Low volatile fatty acids (VFAs) concentration in influent.
- The temperature around the mesophilic ranges (303–311 K)

5.4.1.1 Septic Tanks

Septic tank is the most common and simplest of anaerobic reactors. Quite popular as an in situ sanitation system, these are watertight substructures where discharges from toilets are treated by anaerobic process. It is often used as part of larger sewage treatment plants as the first step in biological treatment. It is not uncommon to find septic tanks being used for treatment of certain industrial effluents too. The processes which take place in a septic tank include sedimentation, floatation, liquefaction of solids, and conversion of organic components into gases. Anaerobic suspended growth is carried out in these reactors. A typical septic tank would be either a single chamber or two chambers depending on the inflow. In a septic tank with two chambers, the first portion sedimentation, floatation, and liquefaction of solids take place. While the sludge settles and undergoes further processing over time, the scum floats up, and the gases are vent off. The second chamber, which receives clear effluent facilitates further reduction of organic load. Because of the conditions inside the tank and the nature of the influent wastewater, substantial methane production does not occur. The gases produced are generally vented off. The effluent from the septic tank is fit for natural water cycle using soil infiltration by means of a soak pit or a dispersion trench. The minimum distance of the soak pit from a water source is determined by the permeability of the soil although there are thumb rules in this regard. Most of the codes across the world for household sanitation indicate it as a viable in situ sanitation method. Septic tanks can be used as a first line of treatment of sewage in sewage treatment plants where separate lines are available for sewage and sullage.

5.4.1.2 Anaerobic Baffled Reactor

Anaerobic baffled reactor (ABR) is an improved version of a septic tank with more than two chambers created using a series of baffles. Baffles are arranged to carry out the flow in an upwards fashion along with all chambers. There will be at least one sedimentation chamber and two to five up-flow chambers. The up-flow will increase the contact time with the active biomass (sludge), resulting in improved treatment. The majority of settleable solids are removed in a first sedimentation chamber (Figure 5.5). Gaseous by-products have to be collected and vented from all chambers. Integrated settler cum baffled reactor (ISBR) is a popular type of anaerobic baffled reactor widely used in hostels, hospitals, flats, etc.

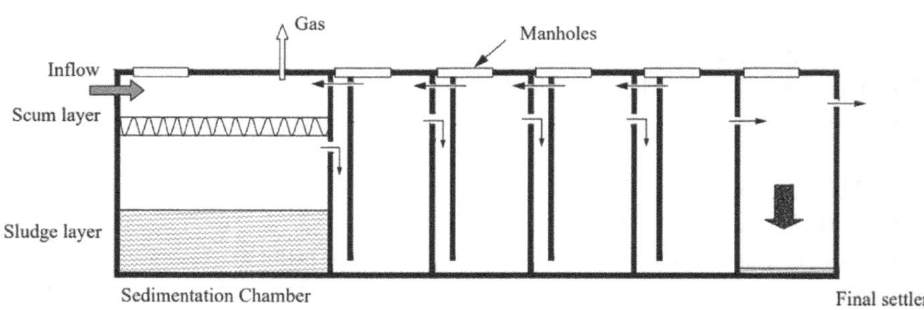

FIGURE 5.5 Anaerobic baffled reactor for secondary treatment

5.4.1.3 Anaerobic Filter

Anaerobic filter was introduced by Young and McCarty in the late 1960s to treat dilute soluble organic waste. Anaerobic filters are widely used as a secondary treatment in households for black or greywater systems and to improve the solid removal compared to septic tanks or anaerobic baffled reactors and sometimes for additional treatment of effluents from a Septic tank or ABR. This unit utilises the anaerobic attached growth process. The filter will be filled with inert rocks or plastic media. The whole filter volume will remain submerged completely. Anaerobic microorganisms accumulate in the media. The biomass within the interstices forms bigger flocs of granular shape due to rising of gas bubble and liquid. This biomass contributes significantly to wastewater treatment. Dissolved and non-settleable solids are removed by anaerobic digestion through close contact with bacteria attached to the filter media.

In single chamber up-flow anaerobic filters (UAF), wastewater flows through the filter bed from bottom to top. When the wastewater comes in contact with the biomass on the filter it is subjected to anaerobic degradation. The filter media are supported on perforated concrete slabs which will facilitate the upward flow of wastewater through the media volume. Backflushing can be done without disturbing the biofilm to prevent clogging of filter media. The media used will have specific surface area typically 100 m^2/m^3 or above. In double chamber up-flow anaerobic filters, the direction of flow in the first chamber would be downwards and in the second chamber would be upward. A conical sludge collection zone would be provided at the centre. The partition wall would have gaps or holes at the bottom portion for movement of the effluent from the first chamber to the second chamber. In the single chamber UAF, the inlet is at the bottom and outlet at the top while in double-chambered UAF, both the inlet and outlet would be at the top. The level difference between the inlet and the outlet would be the driving force facilitating the flow through the media in the double-chambered UAF (Figures 5.6 and 5.7).

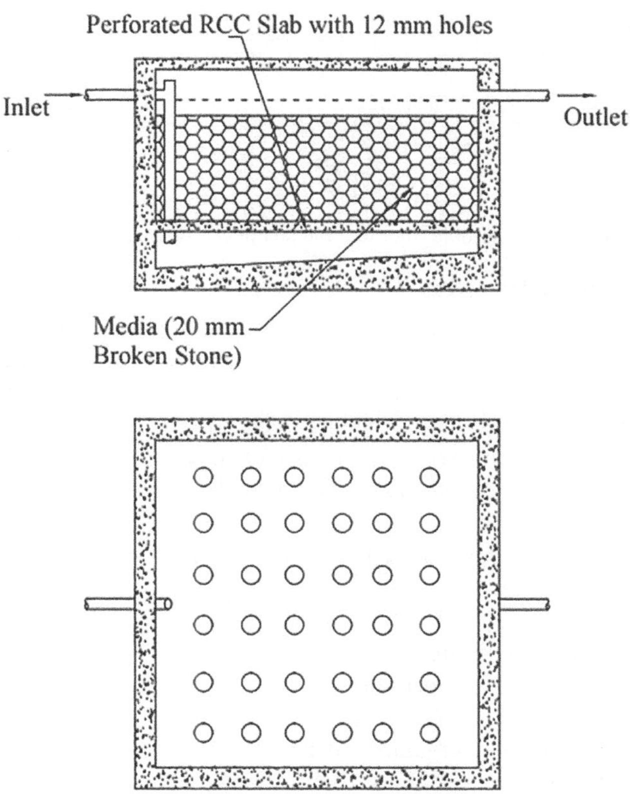

FIGURE 5.6 Up-flow anaerobic filter (single chamber) for anaerobic treatment of sewage

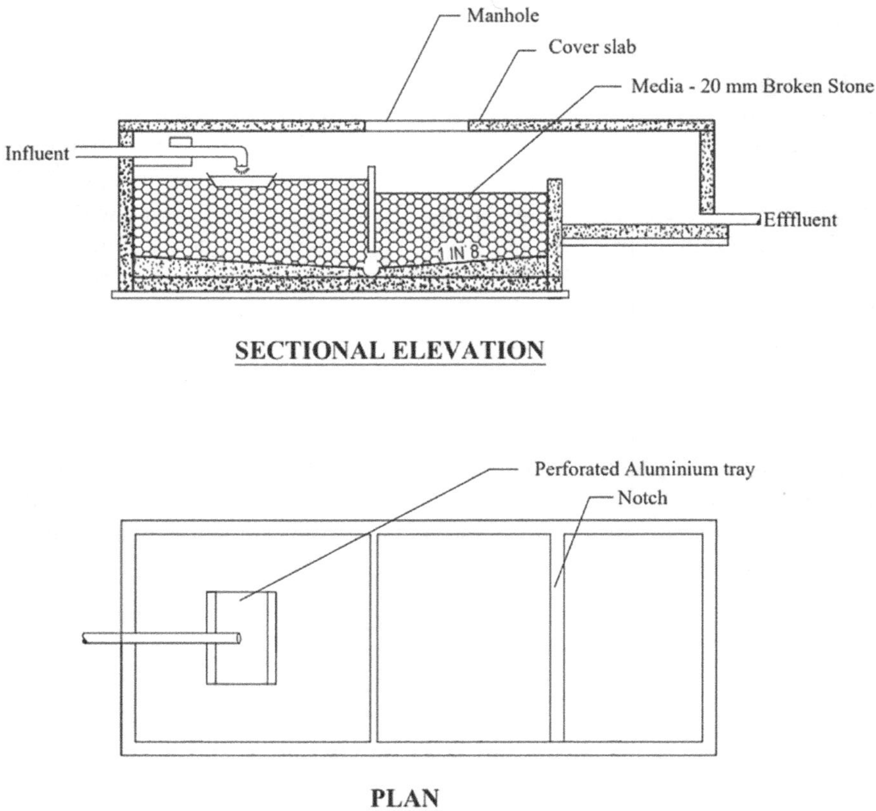

FIGURE 5.7 Up-flow anaerobic filter (double chamber) for anaerobic treatment of sewage

5.4.1.4 Anaerobic Digesters

The combined biological processes of anaerobic decomposition of raw organic matter to soluble organic intermediates and the gasification of the intermediates to carbon dioxide and methane is referred to as the digestion process. Anaerobic digesters provide a suitable condition for the methanogenic bacteria to flourish. Unlike a septic tank, ABR or UAF, an appreciable quantity of methane would be available from Anaerobic Digesters. The methane production is dependent on the substrate (bacterial food) quality.

Conditions that encourage the activity of both acid and methane-forming bacteria include

- an oxygen-free environment,
- a relatively constant temperature,
- a pH between 6.6 and 7.6,
- a consistent supply of organic matter for the bacteria to feed upon,
- high COD, and
- appropriate carbon:nitrogen ratio.

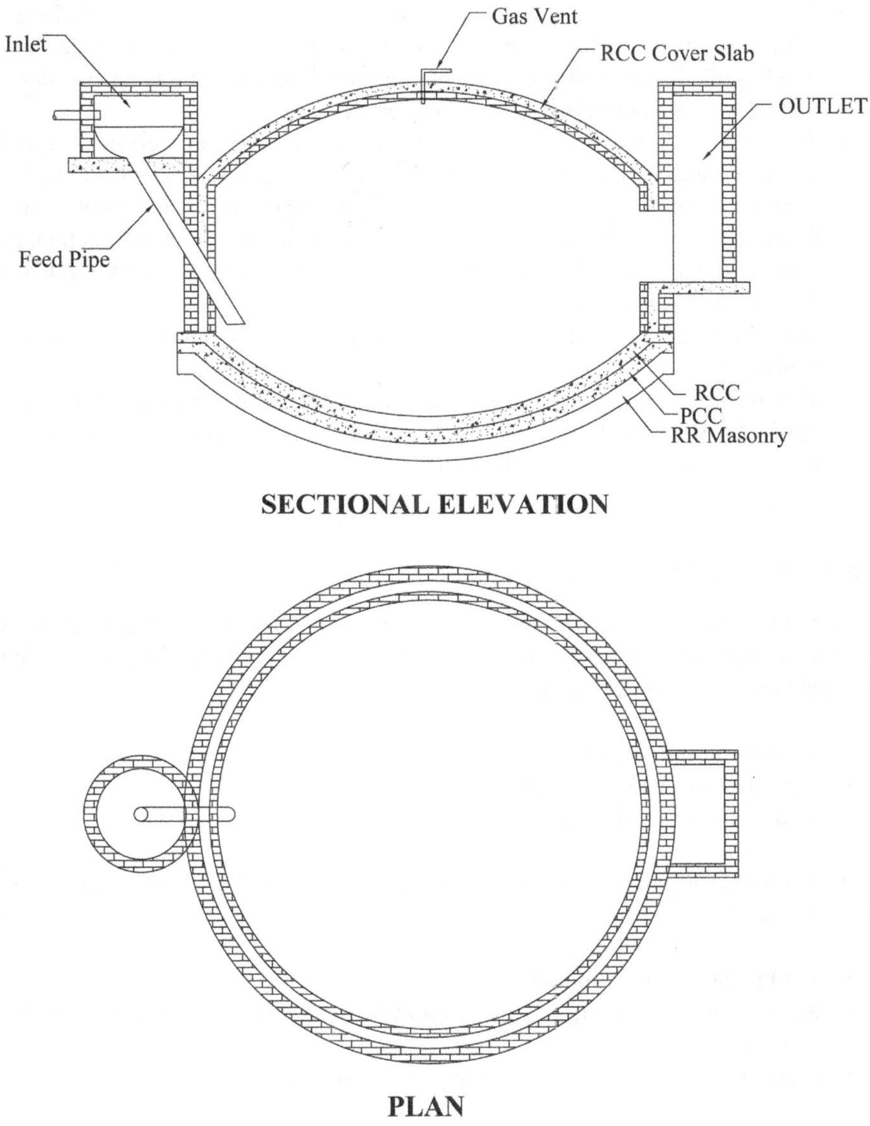

FIGURE 5.8 A model of an anaerobic digester

An anaerobic digester is a sealed tank which provides a suitable environment for naturally occurring anaerobic bacteria to grow, multiply, and convert waste to biogas and a low-odour effluent. The simplest of anaerobic digesters is a biogas plant.

- Typical digesters are rectangular or round concrete tanks with appropriate arrangements for holding the gases produced. The anaerobic digestion process takes place inside these tanks. Inlet arrangements are provided for seeding, feeding wastes, and for manual mixing purpose. Quite often,

degradable solid wastes, agricultural wastes, cow dung, etc., are fed along with raw sewage in biogas plants. Solid portion getting settled inside the tank is called sludge. Sludge from biodigesters has minimum nutrient value and hence is disposed just as soil conditioners.
- The methane gas will be stored under the tank cover. The cover is usually of dome shape in order to store a maximum volume of gas generated. Tank cover comes in two variations called floating dome and fixed dome. In floating dome type the cover will rise according to the amount of methane generated and stored. Collection pipes are attached to the cover to collect the methane generated.
- The slurry will flow to a collection chamber by the pressure of gases generated.
- The most common type of biodigester used are fixed dome bioreactor. The particular shell shape of tanks is for providing minimum area and maximum structural stability (Figure 5.8).

5.4.2 Aerobic Reactors

Aerobic microbe requires free atmospheric oxygen to live and multiply. They use oxygen as an electron acceptor for energy synthesis. Aerobic bacteria mainly help in the oxidation of oxygen-consuming organics (BOD) such as

- proteins (amino acids),
- carbohydrates (starch, sugar, and cellulose), and
- lipids (fats, oil, and grease).

In these biological reactors, oxygen supply has to be maintained. Thus, an aerobic process tank should normally have:

- aerobic bacterial community,
- wastewater input with organic components and nutrients for microbial growth, and
- adequate air supply for the aerobic process to occur.

The aerobic process tanks are also called as aeration tanks. The dissolved oxygen (DO) of water in the tank depends on the system used. It should be in the range of 1.5–3 mg/L for activated sludge process (ASP) and for technologies like membrane bioreactor (MBR) or moving bed bioreactor (MBBR) it will be upward of 4 mg/L and up to 8 mg/L. The bacteria will feed onto the organic material and build up colonies. These gelatinous colonies are called biological floc. These heavy flocs will be later on separated by gravity settling outside the aeration tank. The secondary sedimentation tank (SST) succeeding this aeration tank will separate out this biological floc. Under favourable conditions, the microbial community will flourish and outgrow the available nutrient load. Or else due to inconsistent loading (incoming BOD) or toxic compounds the amount of bacterial population and the microbial health in aeration tank can decline. Then we will have to restore the microbial population by adding bacterial strains externally. Rather than increasing the microbial population in the tank, our objective is to remove out the maximum amount of biodegradable organics with minimum operational and maintenance cost. Keeping this in mind, various aerobic technologies have been invented:

- The systems developed initially were suspended aerobic growth process.
- Then came attached growth aerobic processes. First introduced systems were submerged attached systems with fixed media units.
- Later on, hybrid of attached and suspended systems where the bacterial flocs grow over a media surface and the media is kept in suspension were introduced.
- Recent researches focus on elimination of the succeeding units after secondary treatments. Compact treatment units where the secondary and tertiary level of treatments is simultaneously achieved are now popular.

5.4.2.1 Activated Sludge Process

The activated sludge process was developed in 1914 by Arden and Lockett. The process takes place in two tanks.

- An aeration tank: Here aerobic bacteria feed on organic matters and form flocs.
- Secondary sedimentation: Following a period of contact between the wastewater and the microbes, the outflow is separated from the biological sludge in a secondary settlement tank.
- Sludge recirculation arrangement: The secondary sedimentation tank is provided with sludge recirculation arrangements. Part of the sedimented bacterial sludge is fed back to the aeration tank and another part is disposed of.

This arrangement is to ensure there are enough population of healthy bacteria in the aeration tank. The dissolved oxygen is maintained in the range of 1.5–3 mg/L (Figure 5.9).

Important Terms

i. Mixed Liquor: The liquid mixture of influent wastewater stream from primary units and the recycled activated sludge is called mixed liquor.
ii. *Mixed Liquor Suspended Solids* (MLSS): The quantity of Suspended solids present in the mixed liquor is called MLSS. MLSS concentrations are measured and used by operators to monitor the biomass levels present in the aeration tank. It is an indirect indication of the amount of microbial population inside the aeration tank. In the laboratory, MLSS is also measured similar to other suspended solids test using gravimetric methods. This is a crude method of monitoring the microorganisms present in the system. However, the test does not distinguish between the organic and inorganic fraction of the solids. The convenience of the test, however, generally outweighs this disadvantage.

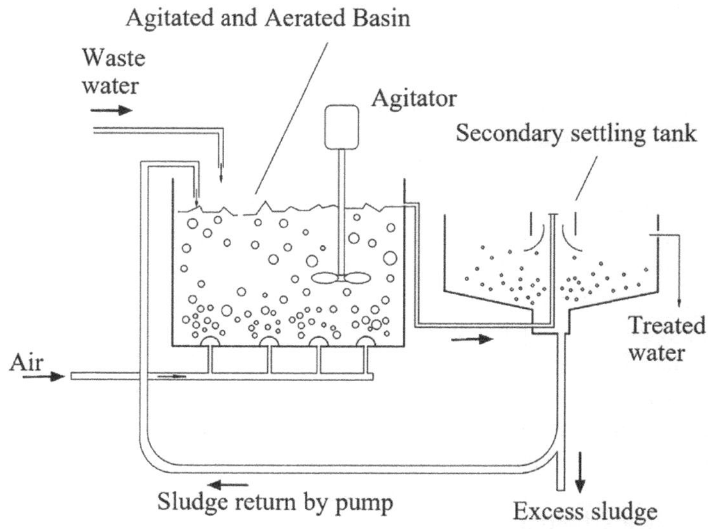

FIGURE 5.9 Activated sludge process configuration for aerobic treatment of wastewater

iii. *Mixed Liquor Volatile Suspended Solids* (MLVSS): The organic fraction can be analysed by burning off the organic contents of the filtered sample at 773 ± 50 K. This fraction is called as mixed liquor volatile suspended solids (MLVSS). Similar to MLSS, it is also expressed in mg/L.

iv. Food to microorganism ratio F/M: The solids loading rate or the food to microorganism ratio is defined as the ratio between the mass of food (BOD) entering the treatment plant and the mass of microorganisms in the aeration tank:

$$ {}^{F}\!/_{M} = \frac{\text{kg BOD / day}}{\text{kg MLSS}} \tag{5.1} $$

If the optimum F/M ratio for the reactor has been fixed, the optimum, or desired mass of MLSS may be calculated by rearranging the above equation. If the actual mass of MLSS is less than the desired level then the concentration of MLSS must be allowed to increase by recirculating more sludge into the system. If the actual mass of MLSS is greater than the desired level then the proportion of wasted sludge from the secondary sedimentation tank (SST) has to be increased.

v. Sludge volume index (SVI): Sludge volume index is an indication of the sludge settleability of the bacterial sludge in the final clarifier

- SVI is the volume of settled sludge in millilitres occupied by 1 g of dry sludge solids after 30 minutes of settling in a 1000 mL graduated cylinder.
- A litre of mixed liquor sample is collected at or near the outlet of the aeration tank itself. This is allowed to settle for 30 minutes in a 1 L graduated cylinder, and the volume occupied by the sludge is reported as:

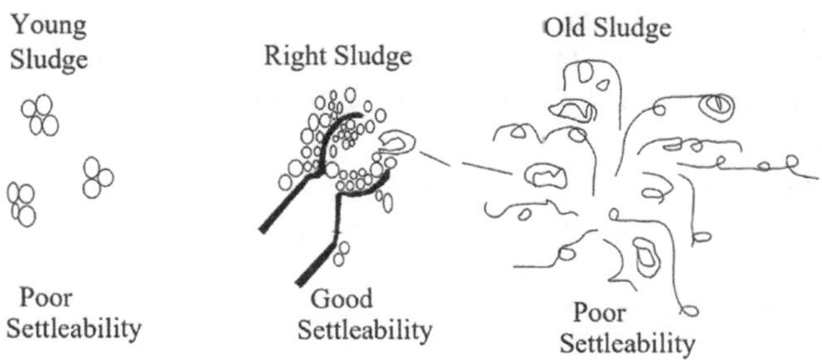

FIGURE 5.10 Representation of sludge settleability

$$SVI = \frac{\text{Settled sludge volme mL/L}}{\text{MLSS mg/L}} \tag{5.2}$$

Ideally, low F/M ratio is preferred for better efficiency of the aeration tank. But a range is fixed considering the efficiency of succeeding units. Low F/M may lead to higher SVI.

Low SVI indicates better settling properties which are desirable for the efficient operation of the ASP system (Figure 5.10).

The settleability reduces due to the appearance of certain types of filamentous bacteria in older sludge. Biological sludge with these bacterial strains is difficult to get separated from the supernatant using gravity sedimentation in SST. So, for deciding the recirculation proportion, SVI is also considered in the equation. Thus, the amount of sludge to be recirculated compared to the total sludge amount in the final clarifier is:

$$\text{Recirculation ratio}\left(\frac{Q_r}{Q}\right) = \frac{\text{MLSS in mg/L}}{\left(\frac{10^6}{\text{SVI}} - \text{MLSS in}\frac{\text{mg}}{\text{L}}\right)} = \frac{100}{\left(\frac{100}{\text{MLSS in}\frac{\text{mg}}{\text{L}} \times \text{SVI}}\right) - 1} \tag{5.3}$$

Determination of Sludge Recirculation Ratio

Method-1 Field Determination Test for MLSS

In the strict sense, MLSS is a gravimetric unit – mg/L and the normal design level is between 3500 and 4000 mg/L in the aeration tank. However, in the field, if the operator does not have ready access to an electronic weighing machine, volumetric measurement using a 1-L measuring cylinder (or jar) is usually performed.

- Take 1 L of the aeration tank sample (the mixed liquor).
- Allow settling in the jar for 30 minutes.
- At the end of the 30 minutes, measure the volume occupied by the settled sludge.
- If it is 350 mL, we take the MLSS to be 3500 mg/L. If it is 400 mL, then take MLSS to be 4000 mg/L.

The assumption here is that the STP is functioning normally, and therefore the Sludge Volume Index (SVI) is 100. Dry solids weighing 1 g occupy 100 mL volume after 30 minutes of settling. So, 4 g of microbes (4000 mg) will occupy 400 mL volume in the cylinder.

For example, to maintaining mixed liquor solids concentration of 0.3% (3000 mg/L), the percentage of sludge that must be returned is:

(SVI =100 as per assumption)

$$Q_r/Q = \frac{100}{\frac{100}{\text{MLSS in mg/L} \times \text{SVI}} - 1} = \frac{100}{\frac{100}{\left(100/0.3\right) \times 100} - 1} = 43\% \tag{5.4}$$

Method-2: Using Calibration Graph

Calibration charts are prepared in plants where fluctuations in load are observed. We have to do the field jar test. With known MLSS and SVI value, the sludge recirculation ratio can be found out. A sample calibration graph is shown below (Figure 5.11):

If the STP is operated within an MLSS band of say 3500 mg/L (350 mL) and 4500 mg/L (450 mL), when the MLSS exceeds 450 mL, the excess sludge is taken out of the system to bring the MLSS down to the say 350 mL, and the process continues until the sludge again builds up to 450 mL. Normally STP should be operated in a smaller band within the allowable MLSS limits.

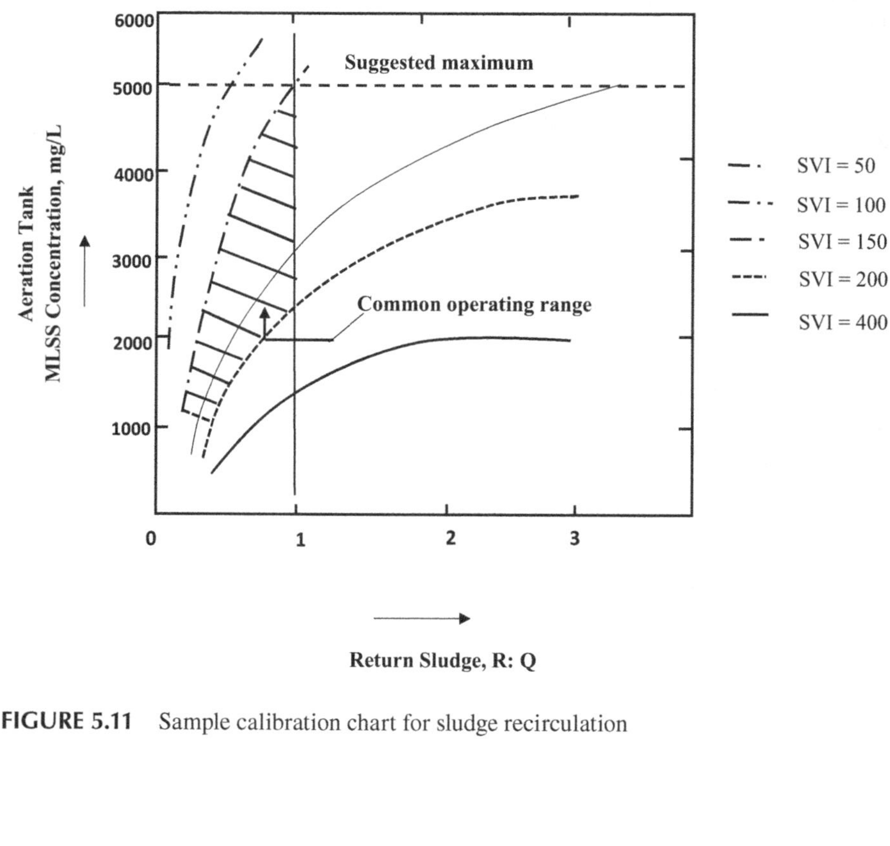

FIGURE 5.11 Sample calibration chart for sludge recirculation

5.4.2.2 Sequencing/Sequential Batch Reactors (SBR)

This is a variation of ASP where aeration and sedimentation process are carried out in a single tank itself. The process of recirculation of sludge is eliminated in this system. Various series of operations similar to that in ASP are applied to a batch of wastewater in the same tank (Figure 5.12).

i. **Fill phase:** First, the aeration tank with bacterial strain and aeration arrangements is filled with wastewater. This phase is carried out providing minimum aeration sufficient enough to mix the bacterial strain and the incoming wastewater.
ii. **Aeration phase:** In this phase aeration is provided for the bacterial action to take place. Like the aeration tank in ASP, the bacterial flocs will be formed in this stage, and the bacteria will assimilate or consume the BOD load. High-efficiency fine bubble non-clog membrane diffused aeration is preferred in SBR tank.
iii. **Sedimentation phase:** Aerators are turned off so that the tank volume is left undisturbed. This provides a condition similar to that of the final sedimentation tank. The bacterial floc will settle down under gravity.
iv. **Decant phase:** Then the supernatant will be separated into the succeeding tertiary units.
v. **Idle phase:** The reactor tank is kept idle for some amount of time. A portion of the sludge accumulated will be removed from the tank. The remaining sludge act as the microbial seed for the next batch.

As the process treats a batch at a time, parallel SBR units are to be provided. This gives benefits where the hydraulic loading has huge fluctuations. During lean flow, we can operate the system with a lesser number of SBR units.

Advantages of Using SBR

- The bacterial population is not disturbed by recirculation devices. Thus, bacterial health is maintained.
- The operation and maintenance cost are low, as it avoids pumps for sludge recirculation.

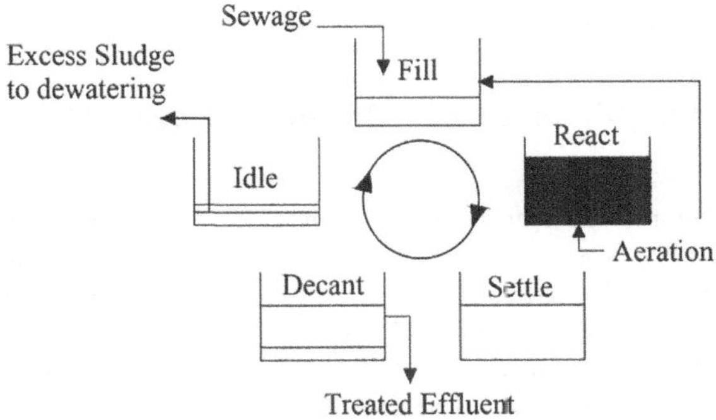

FIGURE 5.12 Sequential batch reactor configuration

- Hydraulic load fluctuations can be accommodated.
- As the tanks are frequently emptied odour troubles and formation of corrosive gases, as in ASP are minimised (CO_2 produced by aerobic reaction will produce corrosive carbonic acids).
- Less manpower only required due to automatic control and easiness to operate and maintain.
- High-quality effluent for reuse is obtained.
- The system can self-generate a stabilised sludge.

5.4.2.3 Submerged Packed Bed Reactor (SPBR)

These are aerobic units where the media are introduced for the bacterial film to attach and grow.

This technology incorporated use of heavy media (density greater than that of water, >(1000 kg/m^3) for the biofilm to attach and grow. Since the density of media is greater than that of water, they will remain at the bottom of the tank itself. Numerous such heavy media are inserted in an SPBR tank. SPBR is not that much in use now.

5.4.2.4 Submerged Aerobic Fixed Film (SAFF) Process

In this process, corrugated inert UV-stabilised PVC media act as a platform for the microbial community to attach and grow. The SAFF media comes as a set of corrugated sheets stuck together. The corrugations will provide more surface area for the biological activities. SAFF technology is in wide use now.

Features of SAFF Technology

i. Reactors up to 6 m deep are possible which enables reduction of land requirements.
ii. SAFF media provides greater surface area for biofloc to attach and grow than SPBR media due to corrugation on the media surface. Effectively, the tank volume required for SAFF tank will be lower than that of SPBR.
iii. The cross-corrugation pattern of SAFF media ensures that air bubble travels in a zigzag fashion, thus increasing the bubble travel path and thereby increasing the oxygen transfer rates.
iv. This higher surface area helps biomass microbes for rapid digestion of biomass.

TABLE 5.2
Advantages and Disadvantages of SPBR and SAFF over ASP

Advantages	Disadvantages
• Tanks have to be constructed deeper in dimension to treat incoming volume saving area • Takes sudden variations in loads • High MLSS can be maintained as the bacteria flocs are anchored well with the media surface • No sludge recirculation is required to maintain MLSS • Rate of sludge generation is normally lesser as compared to a conventional system	• The media is inserted at a depth. These criteria will cause some troubles in installation • Full tank volume is not utilised for bacterial growth

v. The air for the process is supplied through a mechanical aeration system arrangement which consists of a blower and fine bubble diffusers.
vi. Aeration tank is followed by sedimentation tank where the bacterial floc that has got into the mass of the treated water during aeration is settled out.

These attached growth processes yield better quality treated wastewater than conventional activated sludge process. The observed advantages and disadvantages of these attached growth process as compared to ASP are listed in Table 5.2.

5.4.2.5 Moving Bed Bioreactor (MBBR)

MBBR was invented in the 1980s. It's a combination of suspended growth and attached growth processes. Inside the aeration tank, special plastic carriers act as a surface for biofilm growth. Carrier materials are made of materials with a density close to that of water (1000 kg/m^3) an example is a high-density polyethene (HDPE) with a density of 900 kg/m^3. The media will be smaller and lighter compared to that of SPBR or SAFF. Numerous media are to be added to keep the desired level of microorganisms in the tank. Unlike SPBR and SAFF, the MBBR media will be kept in suspension. Carriers are moved inside the tank through the aeration system. This ensures good contact between the wastewater and the biomass on the carriers. Likewise in ASP technology, the aeration tank is succeeded by a sedimentation tank where the microbial mass is settled out (Figures 5.13 and 5.14).

An efficient carrier is one which has the following properties:

- Provides a maximum surface area for biofilm growth
- Provides good anchorage for the biofilm and thus prevent break up of biofilm during a collision
- Facilitates maximum oxygen contact
- Have proper density so that the media remain inside the tank for aeration.

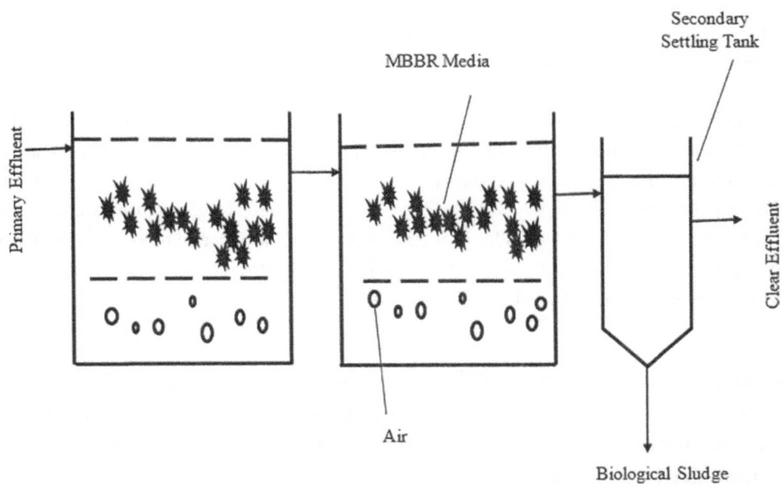

FIGURE 5.13 MBBR configuration

FIGURE 5.14 MBBR media

TABLE 5.3
The Relative Advantages and Disadvantages of MBBR

Advantages	Disadvantages
1. There are no limitations of the height of the tank as in fixed media reactor as finer aeration can be used. 2. The structures can be easily covered for indoor air quality when needed. 3. Requires lower footprints compared to conventional activated sludge. The corrugated media surfaces will provide sufficient area for biological growth. So instead of making the tank bigger, more media has to be inserted. 4. Easy to operate and maintain. 5. Flexibility in handling hydraulic and organic load variations. Extension and expanding are easy. This is made possible by increasing the number of carrier media. 6. MBBR is not affected by toxic shock. 7. MBBR media is installed to retrofit existing activated sludge tanks in order to increase its capacity. 8. Production of sludge is lower. 9. It does not need recycling of the sludge as in activated sludge system (Recirculation line will be provided for use in case there is a condition of low MLSS).	1. Carrier media are patented product. So, they are costly. 2. The area per unit volume of the media offered by various vendors is different and also each vendor advocates his own criteria for the relative ratio of the volume of media to the volume of aeration tank, which makes it difficult to bring about common and validated standard design criteria. 3. The verification of whether the media is moving about the entire volume of the tank or merely clumping at the top layers and if so the method of mixing it up through the tank volume without shearing of the biomass on it are issues of infirmity and which may need gentle movers of the media through the volume of the tank.

MBBR media variations are available. High-quality media with multiple concentric rings and the corrugated surface will give more surface area for the microbial growth. Latest is the swirl media. The efficiency of aeration is enhanced by its design. These media have sharp edges. This will help in bursting the coarser air bubbles near the media surface and thereby helping the aeration to reach into the microbial film.

Aeration of the unit must also be carefully designed. When coarse aeration is supplied, due to the buoyancy of the large bubbles the media may get expelled from the tank. Vigorous agitation by coarse aeration can result in the media shedding the biofilm from the media surface if the media collide with each other or with the walls. The laying of the diffusers and their configuration play a critical role in ensuring proper movement of the media inside the reactor.

So medium type bubble aeration has to be provided. The aeration design will depend on the media used. The DO is to be maintained in the range 4–8 mg/L (Table 5.3).

5.4.2.6 Membrane Bioreactor (MBR)

This technology combines the treatment effect of aeration and secondary clarification in the same tank. Membranes are used to filter out the aerated mixed liquor. Aeration tank and filter membranes are the two components of this treatment technology. Compared to settling in a separate downstream tank, it does yield treated sewage with practically no BOD and suspended solids and hence being clear and virtually transparent.

MBR tanks can sustain mixed liquor suspended solids (MLSS) of three to four times than what is possible in the conventional aeration tanks which in turn offers minimisation of the total footprint of the treatment plant. Diffused aeration is provided in the aeration tank. DO is maintained in the range of 4–8 mg/L. MBR membrane will be interwoven bio-fibre with pores of size in the range of microns or below. Wastewater under pressure will be passed through the membrane and due to incoming force, the water will get filtered out. Membranes are available as flat sheets made of polymeric materials or hollow panels or cylinders. The process configuration can be also of two types (Figure 5.15):

- Membrane immersed in the aeration tank (iMBR)
- Membrane placed outside of the aeration tank as a side stream unit (sMBR)
  - An 'element' can be considered to be the smallest replaceable component of the membrane unit.
  - The term 'module' normally applies to a collection of elements in the case of flat sheets.
  - The term 'stack' can be considered to apply generally to a collection of elements, though the terms 'rack', 'unit', 'cassette' and 'skid' are also used.
  - A collection of immersed stacks is usually called a 'train'.

The membrane is a matter of proprietorship and the throughput per membrane module offered by various vendors are different and also each vendor advocates various shapes of the membranes as flat sheet, cross flow, dead-end flow, etc., which makes it difficult for common validated standard design criteria. Coarse bubble aeration has to be provided in the membrane tank to scour the membrane and keep it clean. Aeration is essential to produce the required shear force for the filtration process to happen. In iMBR fine screens (1–6 mm) has to be provided before the aeration tank inlet. The sludge scored off from the film is fed back to the aeration tank. A part of

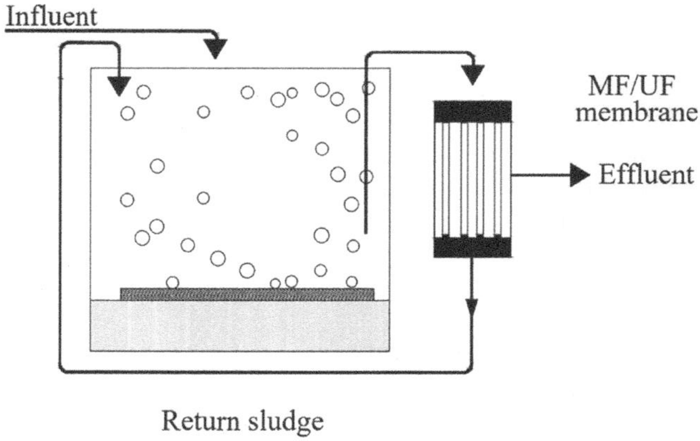

FIGURE 5.15 Membrane bioreactor (MBR) side stream configuration

TABLE 5.4
Comparison of MBR and MBBR

Particulars	MBR	MBBR
Capital investment	Higher	Lower compared to MBR
Requirement of screening	Higher	Lower compared to MBR
Requirement of chemicals	Require membrane conditioners	Require no addition of chemicals
Mechanical devices	• Aerators • air scouring	Aerators
Operation during power failure	Last up to 24 h after a power failure in a healthy reactor	Last up to 10 h after a power failure in a healthy reactor
Effluent quality	Superior quality effluent	Effluent quality sufficient for irrigation reuse

sludge from the aeration tank has to be wasted likewise in ASP. Chemicals called membrane conditioners are also used to scour off the membrane from pore plugging material. MBRs provide a high level of disinfection and also clarify the water so as to increase the effectiveness of a downstream UV disinfection process.

Comparison between Membrane Bioreactor (MBR) and Moving Bed Bioreactor (MBBR) can be done as indicated in Table 5.4.

5.5 AERATION SYSTEMS

Aeration is the process of introducing air into a system. In a sewage treatment plant using aerobic systems, the air needs to mix with the sewage to be treated. There are basically two methods. Either the water is made to come in contact with air as fine droplets or air is diffused into the water. In a typical STP using aerobic process and primary chemical treatment, the units that may use air are the Equalisation Tank and the flocculation tank. The Equalisation Tank uses air to homogenise two different streams and to prevent possible sedimentation. Flocculation tank uses air for chemical mixing and to aid floc formation. For biological processes like ASP, SAFF, SPBR, SAFF, MBBR, and MBR, air is used to meet the oxygen requirement of the process. In ASP and its variants, the process is suspended process where air is used to circulate the biological mass within the reactor also. In MBBR, air supply meets the need to move the media within the reactor over and above meeting the oxygen demands. In MBR, the function of the air includes keeping the membrane unclogged too.

5.5.1 Types of Aeration

Aerators are devices that are used for aeration purposes. There are different types of aerators like surface aerators and submerged aerators. Surface aerators are mechanical systems using blades, propellers, or brushes to agitate the water to introduce air from the atmosphere. The oxygen is brought into the liquid via movement of the liquid surface by which atmospheric air is mixed with water directly. Sprinkled liquid and the air-liquid mixture at the spot of the aerator where the air is forced into the liquid or bubbles that are brought into the liquid by the mechanical agitation. The blade/vanes of the aerator will be at the level of water inside the tank. High-speed agitators are often mounted over fixed platforms. Another type of aerators is mounted over floating platforms so that the aerator will accommodate the level variations inside the tank. Such type of aerators is suitable for equalization tanks where the water level varies with the plant inflow. Based on the axis of rotation of the agitators there are vertical shaft aerators where the moving blades/vanes are suspended from a vertical shaft connected with motor and horizontal shaft aerators in which blades are mounted over a horizontal spindle axis. Motors are required to impart enough mechanical energy to all types of mechanical agitators. Submerged aerators are mechanical agitators where the blades are kept under the water surface.

5.5.2 Air Blowers

Blowers are motor driven mechanical equipment to supply air to the diffusers. These devices suck the atmospheric air and create supply in the air pipe. The power costs of running the air blowers constitute a significant part of the operating cost. Proper selection of air blowers will maximise effectiveness and minimise energy costs. The selection of blowers primarily depends on the volume and pressure required in the aerobic reactor. The capacity of air blowers is generally expressed in m^3/hr. Generally, two types of blowers used in STP are twin lobe blowers and side channel blowers. Air vents have to be provided as a safety measure in all the units where external air supply is provided especially in closed units to remove excess air and gases from various processes to prevent pressure buildup inside the tank.

5.5.2.1 Twin Lobe Blowers

These are positive displacement blowers. Two lobes moving in opposite directions help in creating the fluid flow inside the blowers. They are preferred in conditions where relatively constant flow rate at varying discharge pressures is needed or when small or medium flows are required. A Twin lobe blower consists of a pair of involute profiled (shape of 8) lobes/rotors rotating inside an oval-shaped casing, closed at ends by side plates. One lobe is the driving lobe, which is driven by the external power (motor) while the driven lobe is driven by a pair of gears. Both the lobes thus, rotate at the same speed but in the opposite direction. As they rotate, the air is drawn into the inlet side of the cylinder and forced out through the discharge side.

There is no change in the volume of the air within the machine but it merely displaces the air from the suction end to the discharge end, against the discharge system resistance. Since the lobes run within the casing with finite clearances, no internal lubrication is required. The air, thus, delivered is 100% oil-free. The motor which drives the Air Blower is connected either directly to the blower shaft or by means of a V-belt or gear drive. Nearly every blower installation has at least an inlet or a discharge silencer as a part of the package. Inlet filter is important to keep foreign material out of the system downstream of the blower, due to the system's inability to handle contaminants. There are other considerations too, such as clogging of instrumentation, contamination of the product, erosion of various parts of the blower system, buildup on valves, and other parts.

5.5.2.2 Side Channel Blowers

These are used for applications which require more pressure or vacuum power than a centrifugal blower is able to provide. The compression procedure is absolutely oil-free and provides applications where pollution of the gas is not allowed. Side channel blowers work according to the impulse principle. This means that kinetic energy from a rotating impeller is transferred to the pumping medium and converted into pressure. The motor rated power determines the maximum differential pressure of the blower.

5.5.2.3 Working of a Side-Channel Blower

The impeller is mounted directly on the shaft of the motor and, together with the specially shaped housing, forms the side channel. The pumping medium is sucked in, compressed in the side channel and conveyed to the second stage after one rotation. Here, it is compressed again before being discharged. This way a side channel blower can be used to generate both vacuum and blast air. The two silencers fitted to the inlet and discharge sides ensure quiet operation.

5.5.3 Diffusers

Diffusers are classified based on size of air bubbles (fine bubble and coarse bubble) and design of the diffuser (disc and tube). Diffusers receive pressurised air from air blowers and release the air in to the aerobic reactor.

5.5.3.1 Coarse Bubble Diffusers

Coarse bubbles have more buoyancy and can be provided in deeper tanks. They are provided where mixing has to be provided along with increasing DO amount.

Coarse bubble aeration is produced through orifice arrangements or with porous membrane.

5.5.3.2 Fine Bubble Aeration

In fine bubble diffusers, smaller bubbles result in more bubble surface area per unit volume and greater oxygen transfer exchange. Fine bubble aeration helps to supply dissolved oxygen to the bulk of the volume. Bubbles are produced through porous plates, discs or domes made of ceramic or plastic material, or punched plastic or rubber membranes. There will be an axial pipe running through the floor-length. Lateral pipes are provided at specific distance. Fine pore produces mostly spherical bubbles 1 to 3 mm in diameter. Coarse bubble orifices will be in the range of 3 mm–50 mm. They produce large non-spherical bubbles.

TABLE 5.5
Advantages and Disadvantages of Coarse Bubble Aerators

Advantages	Disadvantages
Less prone to plugging due to air mass coming from the larger hole openings	Consume more energy to operate
Optimal mixing capacity in tanks with thicker mass and sediments	Less efficient oxygen transfer
An ideal solution for secondary treatment phase (If fine bubble diffusers were used in this second phase, floc, sediment and carbonate buildup would narrow or plug its smaller air holes.)	Less contact time – because larger bubbles will ascend more quickly through the wastewater than smaller bubbles

TABLE 5.6
Advantages and Disadvantages of Fine Bubble Aeration

Advantage	Disadvantage
High aeration efficiency	Susceptible to chemical or biological fouling
High oxygen transfer efficiency	More expensive – fine pore diffusers require more routine cleaning and replacement (plus additional costs of maintenance and downtime)
Require less energy to run (compared to coarse bubble diffusers)	Energy cost challenges – when fine pores become clogged, the diffusers may require more energy to operate (than coarse diffusers)
Easily adapt to existing basins (for replacements or upgrades)	Air flow distribution is critical for fine diffuser performance. This requires proper selection of air flow control systems to ensure fine pore diffusers function at peak efficiency levels
Lower volatile organic compound emissions (compared to coarse diffusers or mechanical aeration devices)	Conflicting airflow requirements – aeration tanks may have conflicting needs of airflow: mixing versus oxygen transfer
Satisfy high oxygen demands	Susceptible to chemical or biological fouling

TABLE 5.7
Treatment Plant Units and Aeration Provided

DAF	Pressurised Fine Air
Equalisation	Floating surface aerator/coarse bubble diffusers
ASP	Fixed surface aerators/fine bubble diffusers
SPBR/SAFF	Fine bubble diffusers
MBBR	Coarse/mixed bubble diffusers (need careful designing)

FIGURE 5.16a Air blowers

In MBBR tank, a special spatial arrangement of diffusers is used to meet the air supply demands. The diffusers are arranged in grid format. There will be a central pipe carrying air from blower. Lateral pipes mounted with disc/tube aerators will diverge from this air supply conduit. Medium size air bubbles are desirable in MBBR tanks to keep the media in circular motion. For achieving this, coarse and fine bubble diffusers are provided in a distributed way. Aeration grid can also be utilised to provide tapered aeration. The aeration demand towards inlet may be more than the outlet end in some types of treatment units. In this case, a greater number of orifices will be provided towards the inlet and lesser towards the end (Figures 5.16a–5.16c and Tables 5.5–5.7).

FIGURE 5.16b Fine bubble diffuser operation

FIGURE 5.16c Coarse bubble diffuser operation

5.5.4 Secondary Sedimentation Tank

Although this is a physical separation operation, this step is required only where the secondary treatment is aerobic biological treatment. The discharge from the aerobic reactor is sent to a sedimentation tank to settle the biological sludge under gravity and separate the clear fluids for further processing before disposal and reuse. Since the unit is located after secondary biological reactor, it is called secondary sedimentation tank (SST). The functioning of an SST is similar to that of clarification tank or a PST. However, this tank comes with depth restriction in order to prevent the anaerobic condition in the settled sludge. Parameters like sludge settleability and sludge concentration have to be accounted for while designing these units. The sludge pumps from this unit will have two delivery lines, one to recycle a portion of the sludge to the aeration tanks to maintain MLSS as and when necessary and another line to dispose the excess sludge.

5.5.4.1 Factors Determining the Efficiency of the Sedimentation Process

- Hydraulic loading rate (HLR): This parameter defines the flow velocity per unit floor area of the tank.
- Hydraulic retention time (HRT): It is a measure of average time the wastewater is held in a tank.
- Solids Retention Time (SRT): It is the average time the activated-sludge solids are in the system.

Clarifier performance with floc sedimentation is mainly dependent on the hydraulic loading rate (HLR). The depth and hydraulic retention time are generally considered as second-order factors. The settling efficiency of an SST can be increased by increasing the horizontal settling surface area. By this way, the hydraulic loading rate can be increased, maintaining the same flow rate. This can be attained in different ways. Pioneer method was providing multiple levels inside settling tanks. This arrangement increases the available settling surface area without increasing the tank floor area or depth and thus saving cost. However, such systems can make the removal of sludge difficult due to practical problems. In order to overcome the issues, alternate methods have also been developed. An alternative to facilitate sludge removal was to increase the steepness of the settling unit by which the sludge is forced to go down and get deposited on a single surface. This is the basic concept of the lamella or plate clarifiers.

5.5.4.2 Laminar Clarifiers

These are also called high rate settling units. A stable flow is forced to take place over plates or lamella or inside the tubes where more sedimentation area is available. In a plate settler parallel or inclined plates (lamellas) are introduced inside a settling tank. There are two basic types according to the flow direction: if the water flows in the same direction as the sludge they are called co-current flow and if in the opposite direction it is described as counter-current clarifiers.

Co-current systems: Greatly facilitate sludge removal, but add the problem of the possible creation of water short circuits. In this case, the installation of baffles is required in order to allow water to pass through the plates. Water circulation through the plates can be increased, with the risk of settled sludge resuspension.

Counter-current systems: The collection of the clarified water is produced in the part where the sludge concentration is lower, resulting in better-clarified water. Reduces the risk of particle entrainment. Rising water tends to contact the upper plate, while sludge descends on the lower one. Some disadvantages of this technology are the need for greater lengths to achieve a stable flow regime and higher pressure drop due to greater friction over surfaces.

5.5.4.3 Tube Settlers

This is a system variant in which plates are connected to each other making the flow pathway similar to a conduit. These are called tube settlers. Although the operation fundamentals are the same, tubular units allow working with structurally rigid modules which are lighter in weight. Tube settlers work in two ways. First, they provide a large surface area for the settlement of particles (increase travel distance) and secondly, they are used to accumulate the smaller particles (that escape clarification zone) until they form larger particles that can move down the tubes uniformly. Some disadvantages include sedimentation and sludge deposits on the tube wall causing a reduction in the effective area leading to reduced general performance (Figure 5.17).

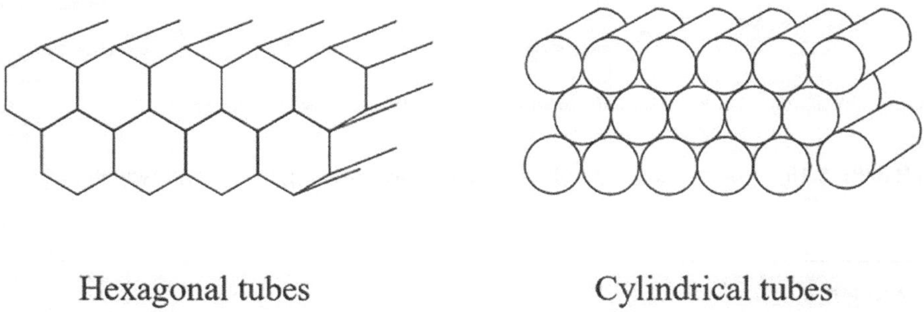

FIGURE 5.17 Tube settler sample arrangements

5.6 COMPARISON OF AEROBIC AND ANAEROBIC TREATMENT

During the anaerobic process, the major portion of waste goes as gas. Thus, the quantity of sludge produced is less and almost stabilised as compared to aerobic sludge. Aerobic systems, use approximately 1 kWh of aeration energy per kg of COD and yields about 50% new sludge from the COD converted, an anaerobic system doesn't require aeration at all and only yields about 3–4% new sludge from the COD converted. In anaerobic effluent the COD will be composed of more unstable mineralised compounds like NH_4^+, PO_4^{3-}, SO_2^-. The effluent has the possibility to be darker in colour and odorous in nature. Because of this reason anaerobic units are always to be succeeded by the aerobic unit. When both anaerobic and aerobic treatments units are located in the same plant, both capital expenditure and operating cost can be reduced (Figure 5.18, Tables 5.8 and 5.9).

TABLE 5.8
Comparison of Aerobic and Anaerobic Treatment

Parameters	Aerobic Treatment	Anaerobic Treatment
Energy consumption	High	Low
Footprint	Low	High
Net sludge yield	Relatively high	Relatively low
Post-treatment	High	Low
Organic removal efficiency	High	High
Effluent quality	Excellent	Moderate to Good
Bioenergy and nutrient recovery	No	Yes
Mode of treatment	Standalone unit or as post-treatment to anaerobic effluent	Essentially pretreatment

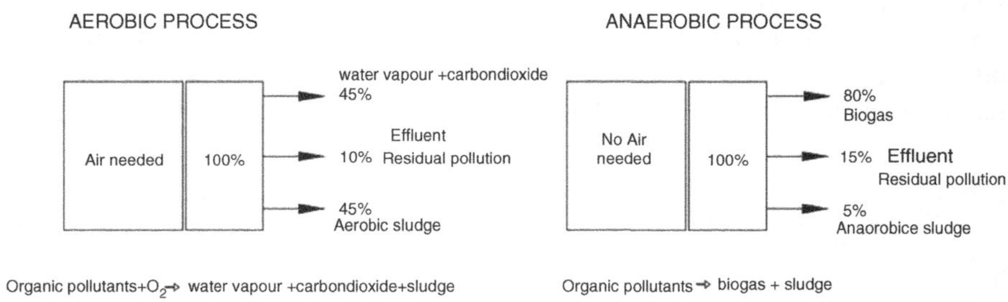

FIGURE 5.18 Comparison of aerobic and anaerobic process pictorial representation

TABLE 5.9
Summary of the Biological Process Discussed

Unit	Process	Process in the Aeration Tank	Additional Arrangements	Succeeding Unit (Effluent)
		Aerobic Treatment		
Activated Sludge Process (ASP)	Suspended growth process	Biological flocs are suspended in the tank aeration tank. Aeration type: coarse bubble/ surface aerator	Secondary sedimentation tank (SST) Sludge recirculation lines from SST to the aeration tank	SST followed by tertiary treatment

(Continued)

TABLE 5.9
(Continued)

Unit	Process	Process in the Aeration Tank	Additional Arrangements	Succeeding Unit (Effluent)
Sequential batch reactor (SBR)	ASP in a single tank	Biological flocs are suspended in the tank aeration tank. Aeration type: coarse bubble/ surface aerator	Multiple SBR tanks in parallel	SST followed by tertiary treatment
Submerged packed bed reactor (SPBR/SAFF)	Attached fixed media process	Biofilm is attached over a static media. Aeration type: coarse bubble/ surface aerator	Submerged media SST Sludge recirculation lines from SST to the aeration tank	SST followed by tertiary treatment
MBBR	Attached suspended media process	Biofilm is attached over a media kept in suspension by aeration. Aeration type: medium type diffused aeration	Floating media SST	SST followed by tertiary treatment
MBR	The aerobic suspended process succeeded with membrane filtration	Biological flocs are suspended in the tank aeration tank. Aeration type: coarse bubble	Membrane unit with coarse bubble aeration. Membrane conditioning agents	Tertiary units
Electrocoagulation	Electrochemical treatment	Electrode arrangements with DC current input	Sludge skimming units in EC-flotation. Anodes need to be replaced frequently	Disinfection
		Anaerobic Treatment		
Septic tank/ anaerobic baffled reactor	Suspended growth process	Sedimentation cum anaerobic reaction tank	Airtight chamber Flow regulation structures Gas venting Sludge accumulation volume	Soak pit/ infiltration trenches Aerobic treatment

(Continued)

TABLE 5.9
(Continued)

Unit	Process	Process in the Aeration Tank	Additional Arrangements	Succeeding Unit (Effluent)
Anaerobic filter	Attached growth process	Anaerobic attached growth over inert media	Pre-sedimentation and screening mechanism Airtight chamber Inert filter media An up-flow mechanism inside filter arrangement Gas vent	Aerobic treatment
Biodigester	Suspended growth process	Anaerobic digestion	Inlet arrangement Gas collection chamber and delivery pipe Slurry collection mechanism	Aerobic treatment Or utilised as plant manure

5.7 ELECTROCOAGULATION

Electrocoagulation (EC) is a treatment technology that removes organic content, suspended solids, heavy metals, emulsified oils, bacteria, and other contaminants from water. The process takes place without the use of any chemicals or microorganisms. Electrocoagulation is performed by applying an electric current across metal plates (electrodes) that are submerged in wastewater. These pairs of metal plates are called electrodes, anodes, and cathodes. Electrode materials will be usually of Iron (Fe) or Aluminium (Al). The cathode loses electrons, and nearby water is reduced to H^+ and OH^- ions. Similarly, metal ions are liberated from the sacrificial anode (Fe^{+3} or Al^{+3}). Heavy metals, organics, and inorganics are primarily held in water by electrical charges. The ions released during the electrochemical process will neutralise the charge of these contaminants and will initiate them to coagulate. The metal ions liberated from anode will combine with OH^- ions and form hydroxide complexes. They act as in situ coagulants. These hydroxides like Fe $(OH)_3$, can remove dissolved pollutants by precipitation or by floatation. The coagulated agglomerates begin to form at the bottom of the tank and can be siphoned out. However, in electrocoagulation-flotation, coagulated particulates would float to the top of the tank by means of formed hydrogen bubbles that are created from the anode. The floated particulates can be skimmed from the top of the tank. EC generally provides a higher degree of treatment than other processes. There is no requirement of biological units, settling tanks/clarifiers or other tertiary treatments units. Generally, disinfection is the only treatment required after EC unit. EC units work under direct current supply. EC modules available in the market are fitted with DC converters. The anode material is getting consumed in the process and would needs regular replacement (Figure 5.19).

The benefits and drawbacks of the EC process are given below.

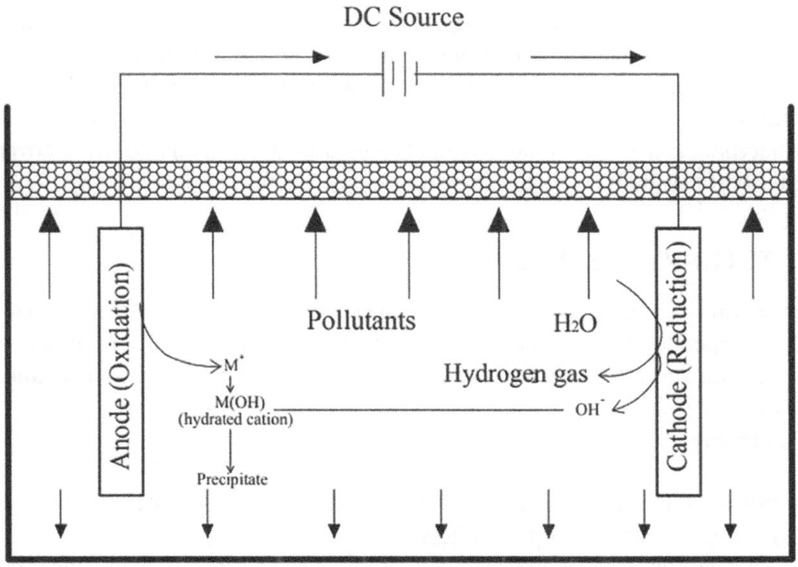

FIGURE 5.19 Electrocoagulation unit representation

5.7.1 Benefits of Electrocoagulation Process

i. EC requires low investment, maintenance, energy, and treatment costs.
ii. EC offers a low footprint treatment system.
iii. EC releases treated wastewater which is colourless, odourless, and clear.
iv. The sludge generation would be minimal.
v. The contaminants getting separated from wastewater which form a surface layer over the EC unit can easily be skimmed off. There are no additional chemicals required in the EC process.
vi. Flocs formed by EC are similar to chemical floc but much larger in size facilitating easy handling.

5.7.2 Disadvantages of Electrocoagulation Process

i. The sacrificial anodes are dissolved into the solution due to oxidation and need to be replaced regularly.
ii. The conductivity of the wastewater treated must be high.
iii. Viscous hydroxide may be likely to solubilise in some cases in which case, the coagulation process would be affected. Some pretreatment would be needed in such cases.

5.8 CONSTRUCTED WETLAND

Constructed wetlands (CWL) are the most popular and well-established natural treatment systems for sewage treatment. It is an engineered mimic of a natural wetland. It is often used as a replacement of the secondary biological reactor or tertiary polishing treatment for effluent from the secondary settling operations. While constructed wetlands are a low-maintenance technology, they occupy much larger land area when compared to mechanised systems but much less than ponds. There are different types of constructed wetlands used for sewage treatment. In principle CWL is a predominantly solar powered aerobic process which has a certain amount of anoxic and anaerobic zones within the system. The details of CWL and NFT are covered in Chapter 6 of this book.

5.9 TERTIARY TREATMENTS

After secondary treatment stage wastewater will still have high microbial load, residual suspended matter, fine particulate content, etc. The tertiary treatment would be to remove these to make the treated wastewater meet the prescribed treatment standards.

Commonly used tertiary treatments are

- Filtration to remove residual suspended matter in wastewater and toxins or pathogens using adsorption filters
- Nutrient removal (N and P) if they are present in excess which is generally quite rare
- Disinfection for removing undesirable microbes and odour

For domestic STP plants, phosphate compounds are removed by coagulation prior to the biological treatment. There is no need for other nutrient removal processes. The effluents from the secondary process are generally filtered through sand media, carbon media and disinfected. According to the secondary treatment technology selected the need for tertiary treatment varies.

5.9.1 Filtration

Filtration is an age-old technology used from ancient times for water purification. Sand and gravel filters were used in India as early as 2000 BCE. Filtration is the process by which solid particles in a suspension are separated utilising their flow through porous media. The porous media generally used in the wastewater treatment are sand or silica minerals like silex, carbon compounds, synthetic fibre, or ceramic media.

Filters are classified based on the rate of filtration and the type of media used. General types according to flow rate are

- **Slow rate filters**: Usually sand filters are working under gravity. The media consist of sand layers. The rate of filtration is very low.
- **Rapid filters**: These are filters using coarser media and the filtration rate will be more compared to slow sand filters.
- **Pressure filters**: Pressure filtration is quite similar to rapid sand filtration. Filtration takes place inside closed chambers under pressurised flow.

For STP pressure filter modules which consume relatively less area and time are used. The contaminant removal mechanism in pressure filters is by straining, settling, and adsorption.

- **Straining**: Straining occurs when particles that are too large to pass through the openings in between media grains are trapped in the filter.
- **Settling/sedimentation**: Settling/sedimentation occurs due to the slowing of a particle's velocity as it passes through the filter. If the particles are slowed sufficiently, they will not have enough velocity to travel through the filter. These particles will get entrapped inside the pore spaces.
- **Adsorption**: Adsorption is the primary mechanism of particulate removal in a pressure filter. In adsorption, particles attach themselves onto the filter media. Although the exact mechanism of the attachment is unknown, it appears likely that there are electrical or chemical attractions between the filter media and the particulate material.

5.9.1.1 Components of a Pressure Filter

- Pressure vessel: Filtration takes place utilising the energy of incoming water which is pumped using a pump or other mechanisms. A pressure vessel is required to conserve this energy so that water is available to move through the filter.
- Underdrain: These are drains provided under the filter beds to collect the filtrate.
- Strainer: These are used to spread water evenly over filter media. They are also used to collect the filtrate. In pressure filters underdrain mainly consists of a strainer and rising pipe.
- Media: sand, silex, anthracite, granulated activated carbon, etc., are used.

According to the type of media used filter can be classified as:

- Single medium (sand/carbon)
- Dual media (carbon and sand)
- Multimedia (carbon, sand, and garnet)

FIGURE 5.20 Pressure sand filter

5.9.1.2 Modes of Operation of a Pressure Filter

- **Filtration**: Effluent after biological treatment and settling is introduced at the top of the media and water gets filtered as it flows down
- **Backwashing**: Good quality water is fed from bottom and the media is washed from impurities. Thus, the filtration efficiency is maintained. The backwashed water is disposed of or fed back to the coagulation tank.
- **Rinse**: This phase is carried out to expel dirt or backwashed water that got trapped between the media. Rinse water fed should be clean and the filtered water should be disposed of. This phase is essential to ensure treatment quality in the succeeding filter phase. In a pressure filter module, these operations are generally controlled by a multi-port valve (MPV). MPV has various settings like filtration phase, backwash phase, rinse phase, and not in use. When we manually set the MPV at the desired function the flow path will be internally modified and desired action is achieved.

Pressure filters come in two variations based on the positioning of MPV.

- **Top-mounted pressure filter**: The MPV is provided over the filter tank. There will be a centre rising pipe running inside the filter tank from top to bottom length.
- **Side-mounted pressure filter**: For taller filter tanks MPV will be connected at the side of the tank detached from filter body. There won't be any rising pipe arrangement.

Operations are the same for both types. But the flow paths in these two types of units are different. Pressure gauges are used to monitor the working efficiency of the filter unit.

5.9.1.3 Filter Media Specification

The efficiency of filtration depends on the filter media properties. When the particle sizes are small, the size of openings available between particles called pore openings will be small. Water seeps down slowly and hence the efficiency of filtration will be better. The rate of filtration will be lesser in these layers. In larger size particles the pore will be greater and the efficiency of filtration will be less.

The particle size of sand is expressed as an effective size (D_{10}). D_{10} is the minimum size of standard sieve (sieve opening in mm) available through which 10% by mass of the sample will pass through. The particle size of fine sand will be in the range of 0.25–0.55 mm for rapid filtration. Apart from particle size, the distribution of particles inside the media is decided by a parameter called the uniformity coefficient (UC). For better filtration, uniformly graded sand layers are required. Similar to D_{10}, D_{60} is the minimum standard sieve size (sieve opening in mm) available through which 60% by mass of the sample will pass through. The uniformity coefficient of a layer is determined as given by Equation 5.5:

$$\text{Uniformity coeifficient} = \frac{D_{60}}{D_{10}} \tag{5.5}$$

Carbon compounds can contribute to removal of organics and odour removal by adsorption mechanism. Activated carbon is typically manufactured from coconut shell or charcoal, the 'activation' process will yield a highly porous material with a very large surface area. Granular activated carbon (GAC) is typically used in Activated carbon filters. Different grades of GAC are available for achieving different reuse standards. The qualities of activated carbon are specified by a term called 'Iodine value' (IV). IV denotes the capacity of carbon material to consume iodine by 100 g of mass. It represents the adsorption capacity of the media.

- Minimum grade to be used is IV 600 (IV stands for iodine value).
- Different grades are IV 600, 900, and 1050 (higher the grade, higher is the adsorption capacity).
- IV 1050 is used for drinking water.

Activated media has to be washed and prepared before installing the unit. While installing the GAC filters are provided after disinfection units. This prevents bacteria from forming colonies on the adsorption sites and thus save the life of media material. All pressure filter units are filled with media only up to 70% of its volume. This is to accommodate the filter bed expansion during backwash operations.

5.9.1.4 Membrane Filtration

Mechanical filtration utilises some type of manufactured media that strains particulate material from the influent water. These types of filters have some type of fabric or polymeric media that has pore sizes small enough to exclude all of the particles larger than the pore size from the filter effluent. By membrane treatment bacteria and other microorganisms, particulate material, micropollutants, and natural organic material, which can impart colour, tastes, and odours to the water and react with disinfectants to form disinfection by-products (DBP) can be eliminated. In STP, membrane filters are the most common mechanical filter found.

- *Feedwater* is the influent water for the membrane system; it is the water being added to the membrane system itself.
- *Flux* is the flow rate through an individual membrane filter module expressed in terms of m^3 of flow per square metre of membrane filter surface area per day.
- *Permeate/product water* is the filtrate from a membrane filter. It is called permeate due to the way that the feed water permeates through the membrane.
- *Concentrate/ Reject* is the waste created from a membrane system. Rejection is the term used to define the percentage solute concentration reduction of the permeate relative to the feed water in a membrane filtration process.

MF, UF and NF use porous media for filtration. UF is carried out using non-porous ion-exchange membrane (Table 5.10).

Air Scrub or AS, is a process by which some membrane systems introduce air into the feed side of the filter module in conjunction with the RO process in order to

TABLE 5.10
Types of Membrane Filtration

Membrane Type	Pore Size	Fate of Contaminant
Microfiltration (MF)	0.1–1 micron	Dissolved solids are passed through. Bacteria and some viruses are removed.
Ultrafiltration (UF)	0.001–0.1 micron Asymmetric or composite membranes	Mineral salts and organic molecules will pass through. Macromolecules, bacteria, and virus are removed.
Nanofiltration (NF)	Around 0.001 micron Used for reverse osmosis	Remove multivalent ions and organic solutes larger than 0.001 micron, bacteria, and viruses.
Reverse osmosis (RO)	<0.001 An asymmetric or composite membrane	Remove dissolved salts, bacteria, and viruses.

enhance the efficiency. The air agitates the surface of the membrane and enhances the removal of the material from the membrane's surface.

5.9.1.5 Filtration Mechanisms

Filtration takes place as water under hydraulic pressure passes through the membrane pores. Filtration occurs by two mechanisms:

- **Straining**: In this mechanism, separation is based on the difference in pore size. Particles larger than the pore size are retained.
- **Diffusion (RO):** In this mechanism, separation is based on the difference in the solubility and diffusivity of materials in the membrane.

Crossflow means that a small portion of the feed water is allowed to flow across the surface of the membrane (rather than through the membrane).

- Crossflow tends to help clean the surface of the membrane and allows for longer operational periods between membrane backwashing or cleaning.
- Crossflow uses slightly more energy than dead-end flow since some of the water being pumped into the module is not passing through the filter.
- Crossflow is usually used on feed waters of lesser quality or if pretreatment with powdered activated carbon is being used.

Dead-end flow means that all of the feedwater entering the membrane is passing through the filter.

Membrane fouling: Deposition and accumulation of constituents in the feed water stream on the membrane will reduce the filtration efficiency. In order to remove any material clogging the filter, a process analogous to backwashing must be conducted on a membrane filter.

Reverse flow is the process of reversing the direction of water flow through the filter using permeate (filtered water). The permeate removes the material deposited on the surface of the membrane and the waste stream is collected and removed from the module, not as effective as traditional backwash procedures. For this reason, membrane filter systems occasionally must undergo a more rigorous chemical cleaning procedure.

Back pulse is a similar process of reversing the flow direction but is used in submersible types of membrane filter systems.

Chemical Clean-in-Place (CIP) is a cleaning procedure that is used to restore the membrane's capacity to something near its original capacity. Each manufacturer has its own unique method to conduct CIPs, but there are similarities in the process. CIP usually consists of allowing the membrane to soak in a relatively mild chemical solution. This solution dissolves much of the material that could not be removed by the RF/AS process and removes it from the membrane's surface. Some manufacturers heat the cleaning solution to enhance its impact. Common methods also require (after the chemical soaking) rinsing the membrane, and then soaking it in a caustic solution. At times, the cleaning solution is allowed to circulate through the filter. These cleaning solutions are collected neutralised and then discarded. The frequency of CIP should be kept to a minimum; a 30-day interval is considered an acceptable CIP frequency.

5.9.1.6 Membrane Module

A single operational unit into which membranes are engineered for use is referred to as a module. It consists of membranes, pressure support structure, feed inlet and concentrated outlet ports, and permeate draw-off points.

Commonly found membrane modules are:

- Tubular module
- Hollow module
- Spiral wound module

5.9.2 Reverse Osmosis (RO)

Reverse Osmosis is a technology that is used to remove a large majority of contaminants from water by pushing the water under pressure through a semi-permeable membrane. Osmosis is a naturally occurring phenomenon where a weaker saline solution will tend to migrate to a strong saline solution. A semi-permeable membrane is a membrane that will allow some atoms or molecules to pass but not others. An RO membrane is a semi-permeable membrane that allows the passage of water molecules but not the majority of dissolved salts, organics and bacteria. In order to make flow happen through a semi-permeable membrane, we need to 'push' the water through the RO membrane by applying pressure that is greater than the naturally occurring osmotic pressure.

5.9.2.1 Working of an RO System

As the feed water enters the RO membrane under pressure (enough pressure to overcome osmotic pressure) the water molecules pass through the semi-permeable membrane and the salts and other contaminants are not allowed to pass and are discharged through the reject stream (also known as the concentrate or brine stream), which goes

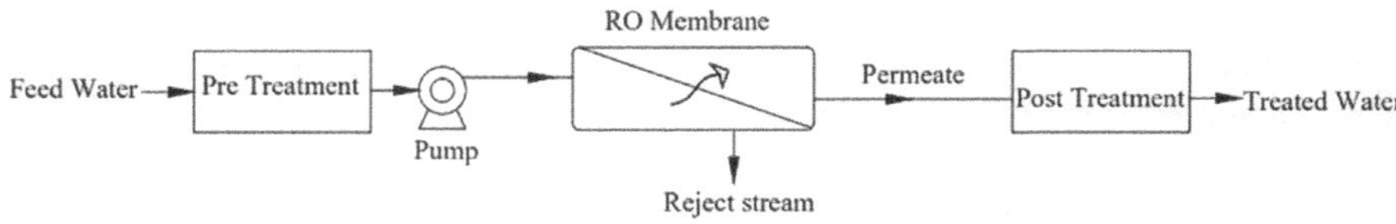

FIGURE 5.21 Schematic diagram of reverse osmosis

to drain or can be fed back into the feed water supply in some circumstances. The water that makes it through the RO membrane is around 95–99% of the dissolved salts removed from it.

5.9.2.2 RO Pretreatments

Proper pretreatment using both mechanical and chemical treatments is critical for an RO system to prevent fouling, scaling, costly premature RO membrane failure and frequent cleaning requirements. A summary of common problems an RO system experiences due to lack of proper pretreatment and the solutions for those problems is given below

Fouling: Fouling occurs when contaminants accumulate on the membrane surface effectively plugging the membrane.

Scaling: When certain dissolved (inorganic) compounds become more concentrated, then scaling can occur if these compounds exceed their solubility limits and precipitate on the membrane surface as scale. An example of a common scale that tends to form on an RO membrane is calcium carbonate ($CaCO_3$).

Chemical attack: Modern thin-film composite membranes are not tolerant of chlorine or chloramines. Oxidizers such as chlorine will 'burn' holes in the membrane pores and can cause irreparable damage. The result of a chemical attack on an RO membrane is a higher permeate flow and a higher salt passage (poorer quality permeate water). This is why microorganism growth on RO membranes tends to foul RO membranes so easily since there is no biocide to prevent its growth.

Mechanical damage: Part of the pretreatment scheme should be pre and post RO system plumbing and controls. If 'hard starts' occur, mechanical damage to the membranes can occur. Likewise, if there is too much back pressure on the RO system then mechanical damage to the RO membranes can also occur. These can be addressed by using variable frequency drive motors to start high-pressure pumps for RO systems and by installing check valve(s) and/or pressure relief valves to prevent excessive back pressure on the RO unit that can cause permanent membrane damage.

Pretreatment Solutions

Below are some pretreatment solutions for RO systems that can help minimise fouling, scaling and chemical attack.

a. Multimedia filtration (MMF): A multimedia filter is used to help prevent fouling of an RO system. A multimedia filter typically contains three layers of media consisting of anthracite coal, sand, and garnet, with a supporting layer of gravel at the bottom. These are the media of choice because of the differences in size and density.

b. Microfiltration (MF): Microfiltration (MF) is effective in removing colloidal and bacteria matter and has a pore size of only 0.1–10 μm. Microfiltration is helpful in reducing the fouling potential for an RO unit. Membrane configuration can vary between manufacturers, but the 'hollow fibre' type is the most commonly used.
c. Antiscalants and scale inhibitors: Antiscalants and scale inhibitors, as their name suggests, are chemicals that can be added to feed water before an RO unit to help reduce the scaling potential of the feed water. Antiscalants and scale inhibitors increase the solubility limits of troublesome inorganic compounds.
d. Softening by ion exchange: A water softener can be used to help prevent scaling in an RO system by exchanging scale-forming ions with no scale-forming ions. As with an MMF unit, it is important to have a 5-μ cartridge filter placed directly after the water softener in the event that the under drains of the softener fail.

   Calcium and magnesium ions (hardness causing) can be removed from water using ion-exchange process. The exchange medium is a synthetic resin, normally in the form of small (0.25–0.5 mm radius) microbeads, usually white or yellowish, fabricated from an organic polymer substrate. They carry a fixed negative charge (functional acid or base group) and are capable of exchanging positively charged ions (cation exchange process) which are mobile ions. The feedwater is pumped from the treated water tank to a tank containing a bed of ion exchange resins. The positively charged ions in the feed water (Ca^{2+}, Mg^{2+}), which has a stronger affinity to the functional group on the resin, are exchanged with positively charged ions available on the resin surface – typically sodium (Na+). When the capacity of the resin is exhausted, it is necessary to regenerate the resin using a saturated solution to restore the capacity of the resin and return the resin to its initial condition. Brine, or sodium chloride solution, is the most commonly used regenerate, although others, such as strong acids (hydrochloric acid, sulfuric acid) or strong bases (sodium hydroxide) may also be used.
e. Sodium bisulphite (SBS) injection: By adding sodium bisulphite (SBS) or Sodium meta bisulphite (SMBS) residual chlorine can be removed.
f. Granular activated carbon (GAC): Activated carbon removes residual chlorine and chloramines by a chemical reaction that involves a transfer of electrons from the surface of the GAC to the residual chlorine or chloramines. The chlorine or chloramines ends up as a chloride ion that is no longer an oxidiser. The disadvantage of using a GAC before the RO unit is that the GAC will remove chlorine quickly at the very top of the GAC bed. This will leave the remainder of the GAC bed without any biocide to kill microorganisms. A GAC bed will absorb organics throughout the bed, which is potential food for bacteria, so eventually, a GAC bed can become a breeding ground for bacteria growth which can pass easily to the RO membranes.

5.10 DISINFECTION

Disinfection is the process of destruction of disease-causing microorganisms or pathogens. It is distinct from sterilisation which involves the complete destruction of all organisms in the liquid being treated. Broadly classified we have chemical disinfection methods and physical disinfection methods.

5.10.1 Chemical Disinfection

Chemical disinfection is the process in which certain chemical compounds (disinfectants) are added to kill the pathogens in treated water. Addition of chemical disinfectants affects the pathogens by causing:

- Damage to cell wall: Either rapture it or alteration of cell permeability
- Alteration of bacterial DNA, RNA
- Inhibition of enzyme activity. (enzymes are secretions from bacteria required for its metabolism)

  Disinfection is carried out inside disinfection tank. The chemical disinfectants are sensitive to chemical and physical parameters inside the tank. Successful disinfection requires selection of proper type of disinfectant according to its working pH and temperature range, a proper disinfectant dose based on desired use and provision of sufficient contact time (time for which water is exposed to a disinfectant and the disinfection is taking place). The volume of the disinfection tank will be based on the contact time required for the disinfection action.

5.10.1.1 Chlorination

Chlorine in gaseous form is the disinfectant widely used during olden days. Nowadays due to its toxicity and inflammability, compounds of chlorine are now utilised for carrying out disinfection. Chlorine is a strong oxidising agent. Disinfection takes place by oxidising the cell enzymes and preventing essential metabolic processes of the bacteria. Chlorine gas is used only in large installations like Municipal Water Supply Schemes due to safety considerations. The safety conditions for a system with gas chlorination needs:

- Exhaust fans below the window sill level
- Ramp for the transport of the cylinder either way
- A lime tank to dump the cylinder in case of leakage, etc.

The most common chlorine compound used for disinfection is hypochlorite compounds. The advantages and disadvantages of using hypochlorite over chlorine gas are (Table 5.11).

TABLE 5.11
Comparison of Chlorine Gas and Hypochlorite

Advantages	Disadvantages
• Inexpensive • Effective at low concentration • Effective in purifying water from pathogens and some inorganic compounds (iron, manganese, and hydrogen sulphide) • Forms residual chlorine	• Chlorine compounds formed after reaction with organics may be odorous, carcinogenic, or mutagenic • Residual chlorine can be toxic to aquatic life

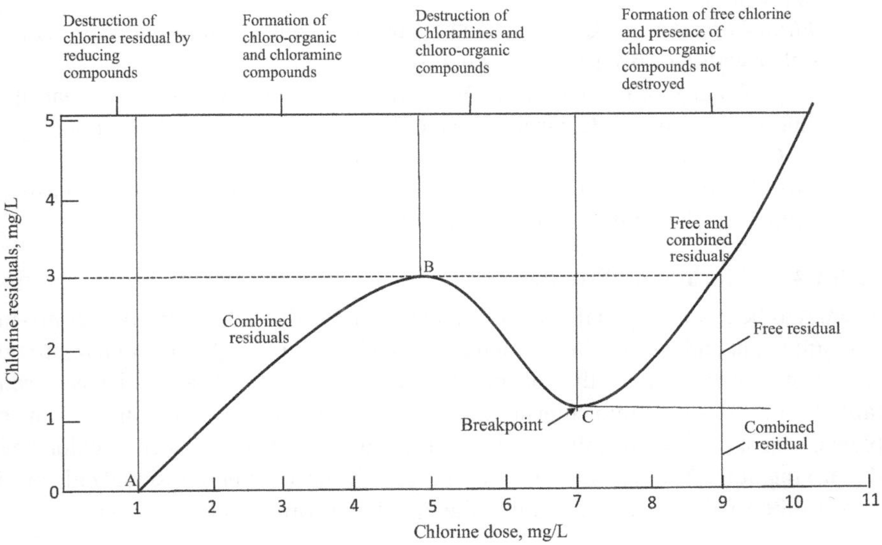

FIGURE 5.22 Chlorine dosing breakpoint

5.10.1.2 Residual Chlorine

When chlorine is being dosed into the water to be disinfected (Figure 5.22):

- Initially, the available elemental chlorine will be utilised to oxidise bacterial cell and enzyme (A-C) in the graph.
- **The breakpoint** is the chlorine dose by which the disinfection demand of the provided water has been met. By this amount of chlorine dosage, the pathogenic will get killed.
- Beyond this dosage, all the extra amount of Chlorine added in the water to be disinfected will appear in the same dose of elemental chlorine.

While disinfecting, the dose of chlorine applied will be according to the residual amount of chlorine permissible for the proposed reuse purpose. In general case, chlorine compound is dosed in an amount where elemental chlorine greater than breakpoint chlorination is supplied.

5.10.1.3 Types of Chlorination

There are different types of chlorination based on the point of treatment it is done:

- *Pre-chlorination*: Chlorine can be introduced into the influent flow to the wastewater treatment plant to:
  ➢ Control odour ➢ Reduce BOD ➢ Aid in settling ➢ Control foaming ➢ Help remove oil
  ➢ Aid in disinfection
  The pre-chlorination before biological units can affect the microorganisms in the aeration tank. Hence residual chlorine is limited in this application.
- *Plain Chlorination*: Chlorine can be introduced within the wastewater plant either within or prior to other processes to:
  ➢ Control or prevent odour ➢ Control or prevent corrosion ➢ Control sludge bulking ➢ Control digester foaming ➢ Control filter ponding ➢ Control filter flies ➢ Aid in sludge thickening
- *Post filtration chlorination:* Chlorine is introduced into the contact basin influent as the final treatment for disinfection

5.10.1.4 Chlorination Process

There will be a tank with baffle arrangements which will facilitate the proper mixing of disinfectant and water. The flow length through the tank will be enough to provide sufficient contact time for the specific disinfectant. There will be a chlorine dosing tank. Using a dosing pump chlorine will be dosed into the tank inlet. In mechanised plants, sensors will be installed to monitor the amount of free elemental chlorine in the mixing tank. According to the reuse purpose if the amount of residual chlorine is to be limited Sulphites are dosed into the tank to perform de-chlorination.

5.10.2 Compounds of Chlorine Used

5.10.2.1 Hypochlorite

Mainly two varieties are used, Calcium hypochlorite ($CaOCl_2$) and Sodium hypochlorite ($NaClO$).

5.10.2.2 Trichloroisocyanuric Acid (TCCA)

This is an excellent source of chlorine with 98% of free chlorine content. The operating temperature ranges are broad. This disinfectant can be used in hot water applications like hot springs and pools. They are also used in industrial and cooling water systems.

They are available in granules or tablets form. Compared to hypochlorite, cyanuric acid compounds are more stable under sun's UV rays. They don't have many side-effects on skin, hair, eye, etc. Because of this property, they are a popular pool disinfectant. Other benefits are:

- Cost-effective and stable source of chlorine
- Easy to handle, ship, store and apply
- Save the expensive cost of dosing equipment
- No white turbidity (as in case of bleaching powder)

The high chlorine content will also provide higher disinfection effect as compared to other chlorine compounds.

5.10.2.3 Bromine Tablets

Bromine tabs are the perfect alternative to chlorine usage in pool or spa. Bromine is as powerful as chlorine but is less irritating to your skin and eyes. Another benefit of bromine is it dissolves at a much slower rate than chlorine. And they don't produce any odour. They don't require the use of stabilisers or other pool water conditioners. Bromine compounds are more stable at higher temperatures than chlorine and more potent at higher pH levels than chlorine.

5.10.3 Physical Methods of Disinfection

This involves using heat, light or sound waves or radiation like ozonation and UV disinfection which are described here.

5.10.3.1 Ozonation

Ozone gas (O_3) can be used as a disinfectant even though it does not leave a residual in the water being treated. It is a highly reactive oxidising agent and rapidly forms free radicals on reaction with water. Ozone gas is generated onsite by the corona discharge process. The process is achieved by applying a high voltage across metal plates. The corona discharge will break the bonds in the oxygen gas. Nascent oxygen will be formed which then combines with O_2 to form ozone gas O_3. Ozone gas is bubbled through fine bubble diffusers. The remaining ozone after disinfection is collected and destroyed either catalytically using metal oxide catalysts or in a furnace heated to 573 K.

Disinfection is typically achieved within five minutes at dosage rates of 5–50 mg O_3/L, depending on the level of treatment received by the wastewater (that is, on how much organic matter is remaining; Table 5.12).

Ozonation is not much used in wastewater treatment. For drinking water purpose ozonation is mostly used for the extra benefit of providing more DO.

TABLE 5.12
Advantages and Disadvantages of Ozonation

Advantages	Disadvantages
• Excellent disinfectants	• The high cost of production
• Oxidation of residual organic compounds	• Potential to cause localised air pollution
• Effective against taste and odour	• No immediate measure of disinfection
• Requires short contact time	• Short contact time reactors prone to short-circuiting
• Biocide property not affected by pH	• Ozone gas is potentially explosive, highly corrosive and toxic
• Removes sulphides	• Highly operational and maintenance-intensive
• Require less space	• More costly than Cl_2
• Contributes DO	• Doesn't have residual
• Ozone can be generated from air, making its supply dependent only upon a source of power.	• May produce harmful by-products

5.10.3.2 UV Disinfection

UV radiation disinfects the water by

- Damaging nucleic acids
- Formation of hydroxyl radicals (strong oxidants)

The effectiveness of UV radiation depends on the dose received by the microorganisms and this depends on

- The intensity of the radiation (the most effective wavelength is 254 nm)
- The path length from the source to the microorganisms
- Contact time at the required dose
- Quality of the wastewater (particularly with regard to turbidity)

Ultra-violet light is produced by a special mercury discharge lamp. Lamps are prone to interference from chemical constituents of the wastewater such as ferric and hardness salts. Periodic cleaning of the lamps is therefore required.

The disinfection property is often reduced by interference due to the absorption of dissolved substances in water and shading of organisms by particulate matters.

UV contractors have very short residence time- seconds to minutes. So, the size of the tank required is very less. In UV contactor tanks, short circuiting is a potential problem which should be considered at the time designing the tank. Usually, the UV lamps are kept either submerged parallel or perpendicular to the flow. Online lamps that are provided over the pipelines are also available. UV disinfection is relatively more expensive than chlorination. Unlike Chlorination, the UV disinfection does not have any residual effect.

5.11 TREATMENT AND DISPOSAL OF SLUDGE

In the context of wastewater treatment residual refers to 'sludge'. Sludge is composed of solids that are settled and separated during wastewater treatment. It is

TABLE 5.13
General Sludge Disposal Methods Used in STP

Solids or Sludge	Disposal
• Screenings	• Sanitary landfill/WTE plants
• Grit	• Sanitary landfill/WTE plants
• Scum/grease	• Transferred to specialist agencies
• Chemical sludge	• Transferred to specialist agencies
• Activated sludge	• Used as soil conditioner/sent to an anaerobic digester
• Aerobically digested sludge	• Used as soil conditioner
• Anaerobically digested sludge	• Used as soil conditioner

necessary to treat properly or dispose of the sludge generated during various stages of wastewater treatment like primary sedimentation, secondary sedimentation, and advanced (tertiary) treatments. The quantity of sludge generated depends upon the quantity of wastewater and the degree of treatment applied. The higher the degree of wastewater treatment, the larger will be the quantity of sludge to be treated and disposed. Properly designed and efficiently operated sludge processing and disposal/reuse system are essential to the overall success of the wastewater treatment plant. The sludge generated during the wastewater treatment can be classified into three categories (Table 5.13).

- **Primary sludge**: Sludge settled in primary settling tanks comes under this category. Primary sludge solids are usually grey in colour, slimy, fairly coarse, and with highly obnoxious odours.
- **Secondary sludge**: This type of sludge from secondary settling tanks has commonly a brownish, flocculent appearance and an earthy odour.
- **Tertiary sludge**: The nature of sludge from the tertiary (advanced) treatment process used.

5.11.1 Dewatering of Sludge

The bulk of the sludge volume, about 90% is water content. Hence, dewatering of sludge is essential to bring it down to a manageable volume. Most of the biological sludge has some nutrient value. Dewatered sludge pressed down to cakes, pellets, or briquettes is often sold as soil conditioners. After increasing its calorific value by mixing with some dry materials sludge pellets are used as fuel too. In the case of chemical sludge, they need to be disposed either by Sanitary Landfill or transferred to specialist agencies dealing with hazardous wastes.

5.11.1.1 Sludge Drying Bed

The water content of the sludge can be reduced to around 70% by this method. The sludge is applied on a bed of sand, which is supported on gravel. A major portion of the liquid drains off in the first few hours after which drying occurs due to evaporation. Sludge cake shrinks, producing cracks which further accelerate evaporation. In dry region generally, the sludge will get dried within two weeks. The drying period will depend on sunshine, rainfall, wind velocity, and relative humidity, apart from sludge characteristics. Under adverse weather condition, it may take up to four weeks. The sludge drying beds should be located at least 100 m away from houses to avoid the smell problem. Sludge drying beds require a large land area (nearly 40% of the total area required for sewage treatment plant).

The construction of sludge drying bed would generally have a graded filter media of coarse sand supported by a gravel matrix and underdrains. The percolate from the sludge drying bed is to be returned to equalisation tank. The dried sludge needs to be removed from the top when it is dried and cracked. Some sand will escape with the dried sludge and over a period of time which needs periodical replacement. Refer Figure 4.5.

The specifications for the preparation of sludge drying beds are as follows:

- **Underdrains**: It is made from the open joined vitrified clay pipe or tiles of at least 0.10 m diameter. The pipe should not be laid more than 6 m apart from each other. The arrangement should be made to return the drained water to primary sedimentation tank.
- **Gravel**: Gravel covers the under-drainage system. Graded gravel is placed around the underdrains in layers up to 0.30 m, with a minimum of 0.15 m above underdrains At least top 30 mm layer of gravel is of 3–6 mm size.
- **Sand**: Sand of effective size 0.5–0.75 mm and uniformity coefficient not greater than 4 is used. The depth of the sand may vary from 0.20 to 0.30 m.
- **Dimensions**: Sludge drying beds are commonly 6–8 m wide and 30 m long. With the bed slope of 0.5%, the length should not exceed 30 m. For the longer length of sludge bed, multiple sludge entry should be used to reduce the length of wet sludge travel. Freeboard of at least 0.4 m is used.
- **Sludge Inlet**: Pipe of minimum 0.20 m diameter should be used as sludge inlet pipe. This pipe should discharge sludge at a minimum height of 0.3 m above the sand bed. Splash plates should be provided at discharge points to spread the sludge uniformly over the bed and prevent erosion of the sand bed.
- **Removal of sludge:** Dried sludge cake is removed by shovel or forks when the moisture content is less than 70%. When the moisture content is less than 40% the sludge is suitable for grinding. Some sand always clings to the bottom of the sludge cake. When the depth of the bed will get reduced to 0.10 m due to this loss, fresh sand of the desired specification is added to match the original depth. After removal of the dried sludge, the sand bed is prepared by levelling for the next cycle of sludge application.

5.11.1.2 Filter Press

At the places where adequate land is not available mechanical dewatering options like filter press is generally used. The plate and frame filter press is the most commonly used method. A high-pressure screw pump would be used to pump the sludge through the filter press. The four main components of a plate and frame filter press are the frame, the filter plates, manifold (piping and valves), and the filter clothes.

Frame: Consist of a thick frame with void centre. Recess is provided in the corner of the frame to drain out the filtered water (filtrate). Usually, detachable types of frame and plate modules are available. In this type sludge enters the first chamber after enough pressure is built up inside and the sludge will move to the succeeding chambers (Figure 5.23).

Filter plates: Consist of a thin frame with filter cloth stretched inside.

An arrangement of frame and plate will create the cavity in to which sludge is pumped, dewatered and sludge cake is formed.

Filter pump: The process happening in this unit is pressure filtration. Sludge is pumped using screw pumps to build enough pressure inside the cavities for dewatering operation. The sludge enters the cavity through an inlet.

FIGURE 5.23 Filter press and screw pump

The phases of filter press operation are the filling phase and the filtration phase. In the filling phase the sludge is pumped into the filter cavity where solids are distributed evenly. In the filtration phase the filter press feed pump builds pressure, solids build within the chambers until they are completely chock-full of solids, forming the cake. The filtrate exits the filter plates through the corner ports into the manifold, yielding clean filtered water. The initial chambers are filled first. Only after the filtration stage is reached in the initial chamber sludge will move to the next chambers. Sludge is kept inside the filter press until the maximum water content is filtered out and the sludge cake takes a dry consistency. The effluent during filter press operation is disposed of into the coagulation tank. Once the chambers are full, the cycle is complete and the filter cakes are ready to be released. After that all the chambers have to be washed to clean the filter cloth for the next batch of operation.

5.11.2 Disposal of Treated Sewage

Treated sewage can be either disposed or reused. Point of disposal can be land or a water body. Reuse can be for various uses like flushing, irrigation of agricultural/horticultural crops, for maintaining ornamental water bodies and sometimes for industrial applications too. In India, it is well known that treated municipal sewage is being sold to thermal power plants for use as cooling water and fertiliser factories as raw water. The treated water so purchased by such factories need further treatment to meet their specifications. However, in case of shortage of water of any type, treated sewage is a life saver for such units. The treated sewage so sold is a continuous revenue stream for the municipal sewage treatment Plant.

For disposal, especially for small communities with a small flow rate, the methods adopted include soak pits and seepage beds. Soak pits are almost the norm in households with in situ sanitation systems with septic tank as the treatment unit. For soak pits and seepage beds, the statutory authorities have developed guidelines with respect to design and criteria for distance from water sources. Soak pits are

TABLE 5.14
Effective Area of Seepage Bed

Percolation Rate mm/h	Specific Effective Area m^2/m^3d
25	50.00
50	25.00
75	17.00
100	12.50
125	10.00
150	8.30
175	7.10
200	6.25
225	5.60
250	5.00

TABLE 5.15
Seepage Bed Dimensions

Bed Dimension	Typical (mm)	Maximum (mm)	Minimum (mm)
Width	1,000–6,000	6,000	1,000
Depth of aggregate	300–600	600	300
Depth of topsoil	100–150	N/A	100
Spacing between beds (sidewall to sidewall)	-	N/A	1,000

constructed with honeycomb masonry structure which facilitate easy radial discharge of the effluents to the surrounding soil. The construction would be such that the Soak Pit is protected from storm water runoffs.

Seepage beds are used to soak the treated sewage into the surrounding soil in situations where soak pits are ineffective due to soil conditions and water table, etc. Seepage beds offer some treatment, albeit small, of the treated sewage prior to soil absorption. A Seepage bed is a bed of aggregate, usually sand, gravel or other inert media, through which the effluent percolates prior to soaking into soil. Treated sewage is applied to the bed through perforated pipes or open jointed pipes laid at predetermined intervals along the bed. Seepage beds are usually rectangular but may be of any convenient shape in plan. Seepage beds should be sized to provide sufficient effective area for the absorption of the average daily flow. The effective area of a seepage bed is the area of the bottom of the bed. The required effective area is determined by multiplying the appropriate specific effective area by the average daily flow (Tables 5.14–5.16).

5.12 SELECTION OF PUMPS

Usually, pumps are provided in pairs. One pump is designated as working and the other is called standby. The rating and specification of these two pumps will be the same. Two methods opted while implementing this system include:

- An alternating or intermittent arrangement, where at pre-arranged times pumps alternate between duty and standby modes, sharing demand time and evening out the running time between them.
- The second approach is a duty/standby arrangement. Here the standby pump is only required in the event of a duty pump failure. The standby pump then runs for as long as needed to repair or replace the duty pump. After that this pump is taken offline and returns to standby mode.
- Standby equipment is intended to minimise downtime (time required for maintenance while the plant operation will be hindered) and keep process systems running.

TABLE 5.16
Common Pumps Used in Sewage Treatment Plants

Unit/Function	Suitable Pump Type
Filter feed pump	Clearwater pump
Sludge collection and transfer	Sludge/slurry pump
RO units depending on the water pressure	Vertical multistage pump/high-pressure pumps
Filter press for pumping sludge	Screw pumps
Conveyance over a large distance where pressure drop is expected	Booster pumps

During the selection of the type of pump, these criteria are considered

- Design requirements
  - The power rating of the motor in kW/hp which is an attribute of head of pumping, volume to be transported and velocity of the flow created.
  - Head of the pump depending on the height fluid has to be pumped.
  - The capacity of pump depending on flow rate requirements at the delivery end.
  - Distance from the suction end and the pumping room: For longer distance (>10 m), the submersible pump has to be used.
- Nature of pumped medium
  The density of the fluid being pumped is an important parameter in deciding pump capacity. For pumping viscous slurry/sludge open impeller pumps have to be used. Similarly closed impeller is used when treated water with less solid content has to be pumped. Based on the nature (corrosion potential) of fluid being pumped the pump material of construction has to be decided. For pumping strong character wastewater Stainless steel (SS) or Cast Iron (CI) body are preferred.

5.12.1 Chemical Dosing Pumps

Chemical dosing pumps have a low flow rate around (0–6 LPH). A dosing pump is a small, positive displacement pump. The dosing pump is fitted either to the cover of the mixing tank (say 50L) or to a wall. Foot valve will be immersed into mixing tank and delivery end will be provided to a contact basin (Figures 5.24 and 5.25).

FIGURE 5.24 Chemical dosing pump

FIGURE 5.25 Chlorine dosing pump with tank

5.12.2 Booster Pumps

Used to increase the pressure of liquid that is already flowing from one place to another in a pipeline. Used in many process applications, where the pressure (head) required for the service is more than one pump can deliver, or where a second pump is required to make sure that the pumps do not cavitate. There can be one (single-stage) or more impellers (multistage) to draw the pumped fluid into the intake of the pump and to boost its pressure as the fluid passes through the impeller and the volute or diffuser casing.

FIGURE 5.26 Raw sewage transfer pump

5.13 SELECTION OF TECHNOLOGY AND DECISION-MAKING PROCESS

5.13.1 Characteristics of the Effluent to Be Treated

The characteristics of raw excreta are almost the same everywhere. But the effluent stream called sewage will have a urine component which influences the characteristics. When it comes to community level sewage, quite often, it would be a mixture of sullage (grey water) and sewage. The presence of sullage can influence the characteristics. In which case, the source needs to be checked for the composition of the two streams if possible or the combined stream. In case of separate streams, the sullage is to be treated to remove inorganic pollutants before combining with the sewage for further treatment. As for sewage or a combined stream with a pretreatment for sullage, the processes basically remain the same. The key factor in deciding on the process with respect to the characteristics include the quantum of nitrogenous material in the wastewater. If the process adopted is aerobic treatment the need for anoxic treatment before an aerobic treatment is dependent on this factor. Even when aerobic process is selected, the system within the group of aerobic technologies to be selected would be influenced by the composition of the raw sewage. In case the context warrants a multi-stage anaerobic treatment, the nitrogen removal would be achieved in those reactors itself which in turn can lead to elimination of anoxic reactors from the process. The pH of the wastewater, if the sullage part has a skewed pH, although not usual, to be treated too will have a significant role in selection of pretreatment methods before the biochemical reactor.

5.13.2 Flow Pattern and Flow Management Options

The flow pattern of the sewage to be treated is of critical importance while designing the treatment plant. There could be sporadic sudden bursts of flow rather than a regular continuous flow in most cases. The 'heart' of a sewage treatment plant is the biochemical reactor, generally designed for a continuous constant rate of flow. In order to manage the flow an Equalisation Tank with adequate storage capacity would be needed. From the Equalisation Tank, by means of a pump with provisions for a return line, a constant rate of flow is maintained to the biochemical reactor. In case the process adopted is Sequential Batch Reactor (SBR), the inflow to the reactor is not continuous and hence there is no need for maintaining a constant flow rate. If gravity feed is possible from the collection tank to the biochemical reactors in the SBR system, the provision for raw sewage pumps can actually be dispensed with. The treated sewage will need to undergo disinfection and filtration. For these purposes too, at least a holding tank would be needed. The disinfection tank generally functions as the Filter Feed Tank. If a higher degree of treatment by means of membrane filtration and reverse osmosis, etc., are needed, there would be more tanks required in the path to manage the flow to such units.

5.13.3 Availability of Power/Source of Power

Availability of power and proximity to a source of power determines the type of technology to be adopted. All aerobic processes and electrocoagulation have several equipment as part of the system which need a source of power in the vicinity and uninterrupted supply of power. Quite often, a generator would be provided in such plants as a backup to meet any emergency situation caused by a power outage. Possibilities of an alternate perennial source of green energy like solar or wind too should be considered at the planning stage of the project. In absence of a good source of continuous supply of electricity, the process to be opted would be a multistage anaerobic treatment and sometimes constructed wetlands subject to availability of adequate land area.

5.13.4 Land Requirements and Financial Considerations

Availability of land has a direct impact on selection of technology. It can be seen that ponds, a completely low maintenance system needing no operators requires very large tracts of land due to the high retention time required. Stabilisation ponds, trendy in the past, are no longer seen as a serious option for municipal sewage treatment due to the large area of land to be committed and the potential for public nuisance like mosquito menace and generation of malodourous gases. If sufficient land is not available, such a large system cannot be considered. Although much smaller in size in comparison to ponds, constructed wetlands too need large tracts of land in comparison with biochemical reactors like anaerobic or aerobic reactors. Within the group of aerobic technologies, the attached growth processes have smaller footprint. Technologies like MBR and electrocoagulation have still smaller footprints occupying a very small area in comparison. It may be of interest to note that smaller the footprint of the technology adopted, higher the Opex would be. For example, electrocoagulation needs the smallest of area in comparison to any other technology but has a very high Opex in terms of both power charges and cost of frequent replacement of the sacrificial electrodes. The choice is between more land to be committed against lower Opex and vice versa. Moving to STPs for apartment complexes, educational institutions, hospitals and shopping malls, smallest of footprints are preferred over other options subject to other factors like Opex.

A set of indicative figures and information on the area requirement and Opex for STPs is given in Chapter 4 Tables 4.14–4.19.

5.13.5 Availability of Skilled Operators

A mechanised system implies need of operators. The operators needed would be either skilled or at least semi-skilled. If such staff are not easily available, the technology to be chosen narrows down to low maintenance systems. Alternately, the Opex has to be budgeted suitably to include the cost of employing skilled or semi-skilled operators from outside.

5.13.6 Mechanism of Operation and Maintenance

When a technology that involves extensive Operation & Maintenance protocols, that involves not only a cost implication but a management system in place to ensure smooth functioning of the plant. The factors involved include budgeting for recurring costs for manpower, maintenance of the equipment, recruitment policy for the personnel, supervisory functions, and allocation of such functions to someone in authority and ownership of the plant.

5.13.7 Climate Impact

While no treatment has a higher impact on climate than in the case of any treatment, the technology with the least implications on climate among the various options need to be selected. However, factors mentioned above too will have to be considered in tandem with climate implications. For example, in cases where land is available in plenty, when there is a choice between mechanised systems and constructed wetlands (CWL), it can be seen that a Constructed Wetland is a better choice from the climate angle since the methane generation in a CWL is negligible. The vegetation in the CWL almost offsets any carbon dioxide generated there. A CWL is almost climate neutral technology. The low maintenance option of anaerobic process releases methane while the aerobic processes release only carbon dioxide which has a lower climate impact than methane. The ecological footprint of a CWL is much lower than any other technology in terms of use of finite resources. While mechanised technologies use power and equipment with moving parts, anaerobic processes have a significant climate impact in terms of the greenhouse gases. It can be seen that CWL is the greenest of technologies available.

5.13.8 Sustainable System for Monitoring

In cases in which monitoring is essential, there should be a mechanism in place for ensuring that. The performance monitoring may need regular analysis in which context a long-term arrangement with an approved laboratory is required. If flow rates are to be monitored, there should be some means of measurement and recording the flows including the human resources required and managed.

5.13.9 Capex and Opex

Capex is of immense importance since in most of the cases in sewage treatment, the Capex would not be recovered. Exceptional cases are observed in water scarce areas where treated sewage is used as cooling water in Thermal Power Plants or raw water for processing in chemical industries. Such situations are possible only if a reasonably large quantity is available, In the case of apartment complexes and shopping malls, treated sewage often get reused for flushing and irrigation purposes. This indirectly has a positive cost implication in terms of reduction in water bills. In such a scenario, the Capex would be a grant based one. An appropriate source to fund the Capex and their budget limitations would a serious factor in selecting a technology. Likewise, the Opex in terms of chemicals, power, cost of human resources, chemicals, spares, and periodic maintenance need to be addressed. There needs to be a source which meets all these costs on a regular basis during the entire life of the project.

5.13.10 End Use

The type of treatment and degree of treatment would be influenced by the end use. The standards for consumption by wild life, agriculture, ornamental ponds, water bodies which function as sources of raw water for drinking purposes need to be met by the treatment plant. The technology selected and degree of treatment would be dependent on the end use as can be seen from the field cases described.

5.14 CASE STUDIES

Sewage treatment plants for three cases with different capacities are discussed here.

5.14.1 STP for a Small Community with 400 KLD with Separate Sewage and Sullage Streams

The scheme is for a small community for which the capacity of the plant is projected at 400 KLD. The raw water is available in two independent streams of sullage.

5.14.1.1 Preliminary and Primary Treatment Units

The preliminary treatment of screening and removal of grit and grease is provided for the sullage stream. The primary treatments for sewage and sullage are done separately.

The sullage is mixed with coagulants like alum, polyelectrolyte, and pH is adjusted by using sodium hydroxide. The flocs so formed are allowed to settle in the primary settling tank. The heavy chemical sludge settles at the bottom of the settling tank while the clear effluent flows out from the top of the settling tank.

In parallel, the sewage undergoes primary treatment in a septic tank followed by a double-chambered up-flow anaerobic filter. The clear effluents from the UAF and the primary settling tank are collected in an equalization tank where the combined stream gets homogenised. The equalization tank is provided with coarse bubble diffusers to supply air from an Air Blower to the effluent in the tank. The aeration keeps bad odours away while helping homogenising the effluent. The Equalisation Tank is a flow management system too which has adequate capacity to hold the combined effluent so it functions as a reservoir from which a steady stream of flow can be ensured to the anoxic tank and the aerobic reactor which follows the equalisation tank. The anoxic tank where the dissolved oxygen is kept at below 1 mg/L removes nitrogen. The aerobic reactor used in this case is a moving bed bioreactor (MBBR). The biological reactor is provided with media with a very large surface area of 600 m^2/m^3. The aeration requirement in the biological reactor is met by means of a set of coarse bubble and fine bubble diffusers connected to an air blower. The treated effluent from the MBBR is sent to a secondary settling tank where the biological sludge settles naturally. This sludge is collected and stored separately. Whenever there is a need to adjust MLSS in the biological reactor, this sludge provides a source of bacteria. The clear effluent from the secondary settling goes through a chemical flocculation and settling to achieve a very high quality of treated effluent with respect to organic load represented by BOD and COD as well as suspended solids. The sludge from this settling tank along with the sludge from the primary settling tank is disposed in a sanitary landfill after dewatering in a sludge drying bed. The percolate from the sludge drying bed is sent to the chemical treatment section for further processing. The biological sludge is sent to the biogas plant where it is processed along with degradable solid wastes in the community.

The clear effluent from the Second Secondary Settling Tank is disinfected and filtered through a pressure sand filter and an activated carbon filter before sending to an artificial lake. The water from the lake is used for irrigation of gardens and lawns (Figure 5.27).

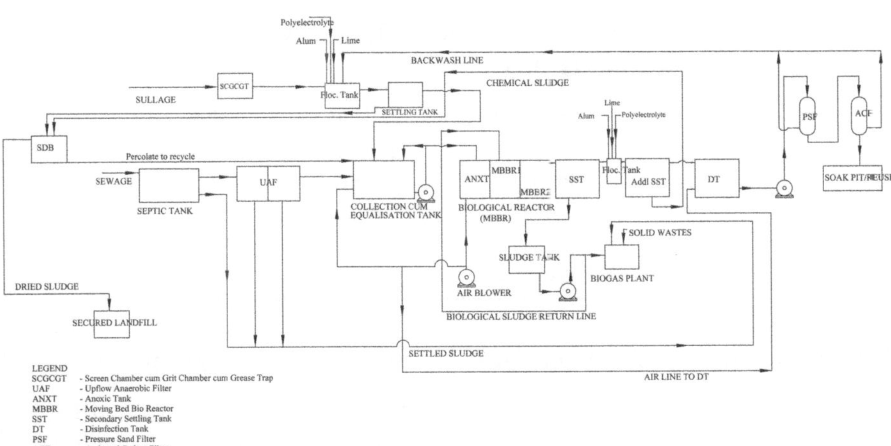

FIGURE 5.27 Sewage treatment plant for 400 KLD flowsheet

5.14.1.2 Treatment Units

The treatment plant has the following units:

- Septic tank.
- Double-chambered up-flow anaerobic filter (UAF).
- Intermediate tank (IMT).
- Transfer pumps.
- Flocculation and chemical treatment unit.
- Chemical solution holders.
- Primary settling tank (PST) for sullage.
- Anoxic reactor.
- Aerobic reactor (MBBR).
- Secondary settling tank.
- Flocculation and chemical treatment unit.
- Second settling tank.
- Chemical doser for chlorination.
- Disinfection cum filter feed tank (DT).
- Filter feed pumps with level controllers in filter feed tank.
- Pressure sand filter (PSF).
- Activated carbon filter (ACF).
- Sludge drying beds (SDB).
- There is a dedicated equipment room built above the underground tanks to house the panel boards, air blowers, chemical storage tanks, chemical dosing pumps, sludge pumps, filter feed pumps, pressure sand filters and activated carbon filters. The picture of the same is attached.
- The results obtained are within the specifications stipulated by the statutory agency.
- The results obtained are given in Table 5.17.

TABLE 5.17
The Characteristics of the Treated Water from the 400 KLD STP

S. No	Parameter	Unit	Value
1	pH	pH units	7.0–7.5
2	$BOD_{(5,\ 20^\circ C)}$	mg/L	<5.0
3	COD	mg/L	<15.0
4	TSS	mg/L	<5.0
5	Oil and grease	mg/L	<1.0

5.14.2 1500 KLD STP Using SBR Technology for a Small Town

This is a small town with a population of about 54,000 of which about one-fifth of the population would be serviced by a sewage treatment plant. The capacity required was project at 1.5 MLD. The treatment plant uses SBR technology for treatment of the

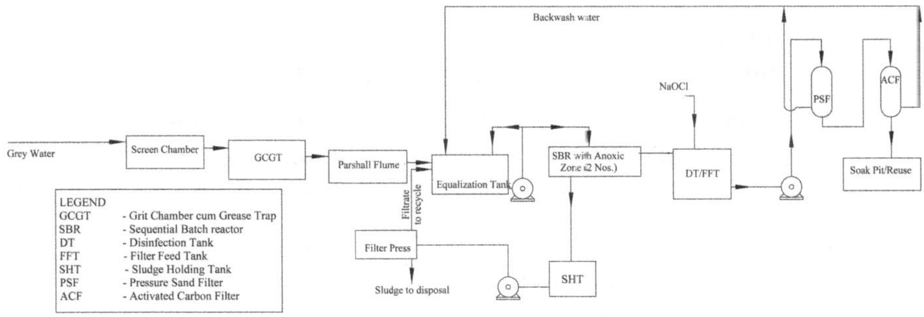

FIGURE 5.28 Sewage treatment plant for 1500 KLD flowsheet

sewage. The sewage collected from the town would be a mixed stream originating from households, commercial buildings, hotels, restaurants, educational institutions and offices, and projected to have the following characteristics.

The selected areas of the town were provided with underground sewer pipes. The beneficiaries were connected to the sewer pipes by means of collection chambers with manhole covers. Due to the layout of the land there was a need for three separate pumping stations within the town to transport the collected sewage to the sewage treatment plant located at the edge of the town. The pumping stations in the town delivers the sewage to the inlet chamber in the STP. The treatment plant consists of the following (Figure 5.28):

- inlet chamber
- screen chamber
- grit chamber cum grease trap
- Parshall flume (as a flow measurement arrangement)
- equalization tank
- SBR tanks with anoxic zones and decanters
- disinfection tank
- pressure sand filter
- activated carbon filter
- treated water holding tank
- sludge holding tank
- sludge dewatering by a plate and frame filter press
- sludge storage facility for complete drying before disposal

The combined stream of sewage gets collected in the equalization tank which is kept in agitated mode always. From the equalization tank, the sewage is transferred to one of the SBR tanks where it undergoes anoxic processing in the anoxic zone and aerobic reaction in the aerobic zone. Since SBR tanks function as the settling tanks too, the sewage is directed to the second SBR tank while the contents of the first tank settles. Once the settling process is complete, the clear effluent is sent to the disinfection tank by means of the decanter. Aeration during the aerobic reaction in the SBR tank is achieved by air blowers. After the aeration cycle, the

TABLE 5.18
Influent Characteristics of 1500 KLD STP

S. No	Parameter	Unit	Value
1	pH	pH units	6.0–9.0
2	$BOD_{(5d,20°C)}$ (max)	mg/L	250
3	COD (max)	mg/L	500
4	TSS (max)	mg/L	400
5	Total nitrogen (as N) (max)	mg/L	40
6	Total phosphorus (as P) (max)	mg/L	7.0

TABLE 5.19
The Characteristics of the Treated Sewage

S. No	Parameter	Unit	Value
1	pH	pH units	6.0–7.5
2	$BOD_{(5d,20°C)}$	mg/L	<10.0
3	COD	mg/L	<50.0
4	TSS	mg/L	<10.0
5	Total nitrogen (as N)	mg/L	<5.0
6	Total phosphorus (as P)	mg/L	<2.0
7	Faecal coliform		<230 MPN/100 mL

biomass settles under perfect settling conditions. Once settled the supernatant is removed from the top using a decanter. Settled solids are wasted from the tank during the decanting phase.

The clear effluent from the SBR tank after the cycle of anoxic reaction, aerobic reaction, and settling is disinfected in the disinfection cum filter feed tank. The disinfected effluent is filtered through a pressure sand filter and an activated carbon filter before disposal. The treated sewage has some application for irrigation and the surplus is be disposed through a series of soak pits (Tables 5.18 and 5.19).

5.14.3 40 KLD STP Using LOMWATS for a Small Community

The project is for a small community of 40 households. The households already had a septic tank and a dispersion trench as an onsite treatment system for the domestic effluents. However, the entire system has become defunct and had to be discarded and new system had to be constructed. The maximum wastewater generation was projected at 40 KLD. Since the plant required is small with limited treatment objectives the treatment required by the community is a Low maintenance Wastewater Treatment System (LOMWATS). The treated wastewater is to be disposed to an ornamental pond in the vicinity. The water in the pond will be used for irrigation of lawns in the area.

5.14.3.1 The Basic Process

The raw effluent after screening and removal of grit and grease in the earlier units, is received in the new septic tank. This flow is by gravity. The effluent from the septic tank is led to the Up-flow Anaerobic Filter (UAF) for further treatment by anaerobic biological process. The discharge from the UAF needs to be mixed with alum, poly-electrolyte, and sodium hydroxide for flocculation and coagulation. This needs to be done using a set of chemical dosing pumps. The flocs formed by the chemical treatment is allowed to settle in a settling tank where the sludge settles down and clear effluent is discharged from the top. The site has a water table which is high. Hence, it was decided that the settling tank will be constructed in a half in half out manner. This in turn required setting up an intermediate pumping station since gravity feed would not be possible. Accordingly, an Intermediate tank is inserted into the system from where the effluent is transferred to the settling tank. The clear effluent from the settling tank is disinfected with sodium hypochlorite solution and filtered through a pressure sand filter and an activated carbon filter before discharging to the ornamental pond in the vicinity.

There will be a fresh water source for back washing of the filters. The back wash water along with the sludge removed from the settling tanks will be dewatered by means of a sludge drying bed the solids from which need to be disposed by means of a Secured Landfill and the percolate to be returned to the chemical treatment tank for further processing. The back wash water from the pressure sand filter and activated carbon filter will be sent to the chemical treatment section for further processing as the same will contain several pollutants.

5.14.3.2 Treatment Units

The treatment plant has the following units.

- septic tank
- double-chambered up-flow anaerobic filter (UAF)
- intermediate tank (IMT)
- transfer pumps
- flocculation and chemical treatment unit
- chemical solution holders
- settling tank (ST)
- chemical dosing for chlorination
- disinfection cum filter feed tank (DT)
- filter feed pumps with level controllers in filter feed tank
- pressure sand filter (PSF)
- activated carbon filter (ACF)
- sludge drying bed (SDB)
- sanitary landfill
- the system yields treated effluent with a $BOD_{(5,\,20°C)}$ of less than 10 mg/L and COD of less than 50 mg/L.

There is a dedicated equipment room built above the underground tanks to house the panel boards, chemical storage tanks, chemical dosing pumps, filter feed pumps, pressure sand filter, and activated carbon filter (Figure 5.29).

The results obtained are within the specifications stipulated by the statutory agency.

The results obtained are given in Table 5.20.

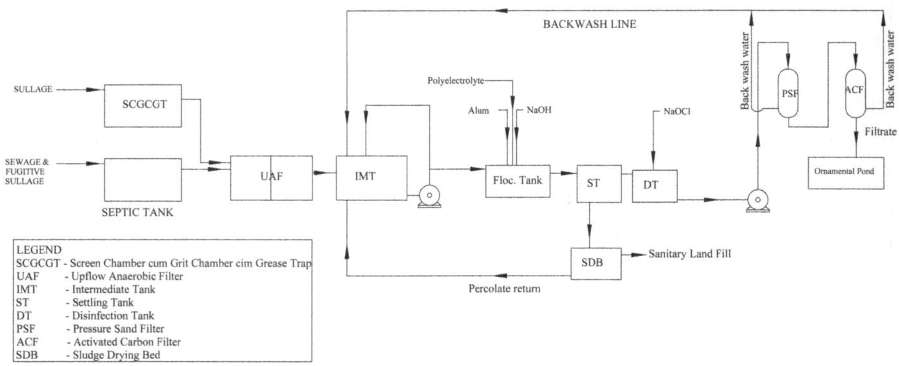

FIGURE 5.29 Schematic flow diagram for 40 KLD STP

TABLE 5.20
The Characteristics of the Treated Sewage

S. No	Parameter	Unit	Value
1.	pH	pH units	7.0–7.5
2.	$BOD_{(5,20°C)}$	mg/L	5.0–10
3.	COD	mg/L	25–40
5	TSS	mg/L	3.0–8.0
6.	Oil & Grease	mg/L	<1.0

6 Grey Water Management

6.1 GREY WATER MANAGEMENT IN VILLAGES AND SMALL TOWNS – INTRODUCTION

In small towns which have a sewer system and a centralised sewage treatment system, all the wastewaters from a town would be drained to it for treatment and disposal. The possibilities and options are described in Chapter 5. However, in smaller towns or towns without sewer systems, grey water management acquires critical importance. Technically, sullage and grey water are the same. However, the term grey water is used here with reference to wastewater flowing in municipal drains. In the absence of sewers and STPs, the sanitation would be in situ systems. The sullage often gets drained into a municipal drain while the in situ sanitation system deals with raw sewage. In the case of in situ sanitation systems, the wastes to be managed would be faecal sludge which had been considered in Chapters 3 and 4. Grey water is generally the wastewater received in municipal drains from various sources within the town. Most of the drains are designed to carry only rain water and storm water runoff to a receiving water body. In practice, domestic sullage as well as discharges from various other sources enter the municipal drains. During rainy seasons, the grey water would get mixed with storm water and the discharge would be a mixture of storm water and grey water. The drains normally drain to a receiving water body. This causes pollution of the water body due to the uncertain characteristics of the grey water. The water body which receives the grey water could be a lake, stream, or river which would be a fresh water source for the same town or some other towns. There could be situations in which the poor quality of water in a lake affects the groundwater in the area which in turn forces the residents in the area to procure fresh water from far off sources.

A general feature of a typical village or a small town in India is dependence of the residents on a water source within or in the periphery of the town. The water source, generally a pond may be receiving water from various types of drains within the town. Traditionally, the drains connected to the ponds were exclusively for carrying only storm water and nothing else. In many old towns, even bigger, with extensive drainage system for storm water too, the storm water drains were not supposed to receive any domestic or commercial wastewater and used to drain into a water body. Effectively, the pond operates like a storm water harvesting system which should sustain the town till the next rains. In some villages, there could be rivers or streams which too would be receiving the storm water through a network of drains. There are villages with more than one pond in which one is a trusted source of fresh water and is used as a source of drinking water while the other which is open to receiving suspect discharges is used only for irrigation purposes. The second pond usually is the one used for ablutions with the first one used strictly as a source of potable water. These classifications were there historically and were made with traditional wisdom and in the light of experience.

DOI: 10.1201/9781003650737-6

In certain parts of India, drains have dedicated purposes. One type of drains will carry just storm water to one or more dedicated water bodies while another set of drains will carry different types of discharges generated by human activities along with storm water to separate water bodies. The water body holding harvested storm water alone is traditionally considered a source of potable water while the water from other water body would be used exclusively for irrigation. This had been a tradition in various parts of India for centuries. (In Tamil Nadu, the term 'Vaari' is used to indicate storm water drains leading to water bodies holding potable water and the term 'Vaayckaal' indicates drains carrying a mixture of storm water and effluents generated by human activity). Notwithstanding such separation in one part of the country historically, quite often drains carrying only mixed effluents can be observed in most parts of the country. The term 'nala' (written as nullah earlier) widely used in North India, carries all kinds of discharges along with storm water. The wastewater in such drains is referred to as grey water.

Due to poor water literacy, the importance of the water body like a lake or a pond in the middle of a village was gradually forgotten. The responsibility of fresh water got outsourced to the government instead of the individuals within the society and the society itself over a period of time. Thus, nowadays, we often come across small towns with a large pond in the middle of the town receiving all kinds of discharges from the human activities there and maintained like a cesspool while the fresh water requirements would be met with by transporting river water from several kilometres away. Traditionally, these ponds were supposed to receive only storm water and the pond was maintained well to ensure a sustainable source of fresh water for the town. It is in this context that grey water management acquires importance.

The grey water management in villages is to ensure that the water body which receives the various types of discharges from the drains, be treated to standards acceptable to the proposed end use. The water bodies could be ponds, lakes, streams, etc. Depending on the classification of the drains, there could be some differences in the approach to treatment. Grey water is likely to be polluted with organic matter and it includes

- faeces,
- discharges from septic tanks/leach pits,
- animal wastes, and
- non-degradable and degradable solid wastes.

The grey water in villages is essentially effluents in village drains. The effluents from the drains to be treated will be having different characteristics, different flow rates and patterns, and require different degree of treatments based on the proposed methods of end uses.

6.2 TYPES OF WASTEWATER

Grey water can be domestic sullage, but can have fugitive effluents of various types. The drains may contain effluents from the following sources:

- Industrial (manufacturing units).
- Commercial (shops, eateries, workshops, etc.).
- Institutional (educational institutions, government offices, etc.).
- Agricultural (run offs from agricultural fields, etc.).
- Domestic sullage (bath water, wastewater from washing of utensils, clothes, and domestic animals).
- Effluents from septic tanks not connected to soak pits or overflow from soak pits and leach pits.
- Raw sewage from toilets not having any kind of substructures, although uncommon, but a possibility in rarest of rare cases.

6.3 CHARACTERISTICS OF WASTEWATER

Characteristics of wastewater can be classified as physical, chemical, and microbiological (Table 6.1).

- Physical characteristics include temperature, colour, turbidity, and odour.
- Chemical characteristics include pH, suspended solids, settleable solids, oil and grease, dissolved chemicals indicated by TDS, chlorides, sulphates, phosphates, nitrates, etc., and organic load indicated by BOD, COD, DO, etc.
- Biological characteristics indicated by microorganisms and pathogenic organisms like *E. coli*, *Salmonella*, and *Helminths*.

TABLE 6.1
Typical Characteristics of Grey Water in Village Drains

S. No	Parameter	Unit	Range
1	SS	mg/L	<500
2	pH	pH Units	6.0–8.0
3	BOD	mg/L	50–100
4	COD	mg/L	200–500
5	Oil and grease	mg/L	<5
6	Total residual chlorine	mg/L	BDL
7	Ammoniacal nitrogen (as N)	mg/L	5.0–10
8	Free ammonia (as NH_3)	mg/L	2.0–5.0
9	Arsenic	mg/L	BDL
10	Mercury	mg/L	BDL
11	Lead	mg/L	BDL
12	Cadmium	mg/L	BDL
13	Total chromium	mg/L	BDL
14	Copper	mg/L	<0.25
15	Nickel	mg/L	<0.25

6.4 TECHNOLOGY SELECTION AND DECISION-MAKING PROCESS FOR GREY WATER TREATMENT

Selection of technology and decision-making process depend on the characteristics of grey water to be treated and the flow pattern. Generally, grey water is expected to be a weak effluent. However, Grey water does not have any standards under any circumstance and the characteristics vary widely from case to case. The main parameters to be considered include suspended solids, pH, BOD, COD, ammoniacal nitrogen, total Kjeldal nitrogen (TKN), and oil and grease. While organic loads represented by BOD, COD and TKN may need biochemical processing, pollutants like total suspended solids (TSS) and oil and grease need physical treatment or a combination of physical and chemical treatment. High level of total dissolved solids (TDS) is not even a remote possibility unless the grey water receives industrial effluents. Industries, irrespective of their size, class and products are expected to have their own effluent treatment plants (ETP) and means of disposal of treated effluents. The flow pattern of the grey water to be treated is of critical importance while designing the treatment plant. There could be sporadic sudden bursts of flow rather than a regular continues flow in most cases. However, there could be a small portion of the effluent which may be entering the water bodies in a continuous manner. The treatment plants, if there is a biochemical process involved, are generally designed for a continuous constant rate of flow. In order to meet this requirement an intermediate tank, often an equalisation tank with adequate storage capacity would be needed. Further, there should be a mechanism, often a pump, to ensure continuous constant rate of flow to treatment system. If the treatment plant is to handle only small quantities of grey water, such facilities can be dispensed with. A higher flow rate would be assumed for designing the plant, and the cost of a higher capacity plant would be less than the cost of having an intermediate equalisation tank and a pump. The presence of a pump in the system would imply availability of power, cost of power, necessity for periodic maintenance, and a cost of replacement over a period of time. Such issues can be eliminated if the treatment plant is overdesigned. In the case of a batch process involving physical separation or a combination of chemical and physical processes, such flow management systems are not needed. Land and power availability, financial availability, operation, and management, skilled labour requirement, climate impact and sustainability are to be considered. These are discussed in Section 4.2 of Chapter 4. The type of treatment and degree of treatment would be influenced by the end use. The standards for consumption by wild life, agriculture, ornamental ponds, water bodies which function as sources of raw water for drinking purposes are to be met while selecting the technology.

6.5 TYPES OF TREATMENT

6.5.1 Primary Treatment

The primary treatment is to reduce pollution loads of suspended matter and non-degradable solid components as well as oil and grease, etc. This is common to all types of effluents to be treated. However, certain steps listed below may not be necessary in treating grey water having just a high level of turbidity.

- **Screening** – to remove large particles in the influent water – mandatory for treatment of drains and nalas carrying any types of discharges. While screens remove the solid part of the raw effluent, disposal of the same is another challenge. Ideally, the screenings are to be handled as solid wastes and disposed suitably as per the options given in Chapter 3.
- **Removal of grit, oil, and grease** – normally done through as grit chamber cum grease trap. The grit, which generally is denser than water, settles to the bottom of the grit chamber while the fats, oil, and grease get trapped at the top of the grease trap which can be skimmed off manually. In specific cases with high oil content, mechanical means like disc or belt skimmers may be needed. Under normal circumstances, such a situation is not expected in the village drains.
- **Coagulation and sedimentation** – May not be necessary in a typical village scenario. If suspended particles do not settle by gravity alone, using chemicals to coagulate and make particles larger and denser, will facilitate settling. There would be a chemical mixing unit by means of a flash mixer or Parshall flume followed by either a clariflocculator or a flocculation unit and a clarifier/sedimentation tank in series. It is similar to the chemical treatment of sullage described in Chapter 5.
- **Sedimentation/floatation** – Again not common in the case of village drains. Instead of making the settleable particles heavier by size agglomeration, this method makes the particles lighter by making them attach to air bubbles so that they float up to the top and be removed. The details of dissolved air floatation are given in Chapter 5.

6.5.2 Additional Options for Primary Treatment

Often, cases where removal of turbidity could be the objective of the treatment of grey water since in many cases, the drains need not necessarily carry organic load at all. In such cases, the options available include

- rock filter,
- roughing filter,
- slow sand filter, and
- rapid gravity sand filter.

All these methods are for reduction of turbidity and suspended solids. There would be some impact on organic load represented by BOD and COD. Similarly, a certain amount of nutrient removal too would be achieved in these filters. Slow sand filters are known to improve the microbiological characteristics significantly. Typical design velocities are given in Table 6.2. Although Pressure Filter (with sand or multi-media) mentioned in Chapter 5 is a type of filter, it is mainly intended as a tertiary treatment after biological treatment. Filters like slow sand filters, rapid gravity sand filters, and pressure filters are generally used only for filtration of drinking water for municipal water treatment and supply.

Rock filters offer very basic treatment and they are not normally considered unless the wastewater to be treated is of very low suspended solids. Rock filters offer a reasonable low maintenance cum low Capex system for treatment of grey water in such a case. While rapid gravity sand filter (RGSF) is an option from a technical perspective, it is not exactly a low maintenance option and hence not considered in a village scenario. A rapid gravity sand filter needs a facility for backwashing which requires involvement of an air blower and a wash water tank at a sufficient elevation to achieve the required pressure. In order to achieve perfect filtration and proper back washing, a RGSF is constructed with an underdrain connected to several laterals. That makes the design and construction a little too complex in village-level context unless the option is for treatment and distribution of drinking water. Further, a RGSF needs at least a semi-skilled operator for stopping the filtration process and resort to backwashing as and when required. The slow sand filter (SSF) is considered an option only when the end use of the water is for potable purposes. All these methods are for reduction of turbidity and suspended solids. But they do remove a certain amount of coliform too although the rock filters and roughing filters per se are not considered a means of removal of coliforms. Slow sand filters are known to remove up to 90% of coliforms too, and hence are an ideal system for a low maintenance drinking water plant. RGSF too has features similar to SSF.

In practice in the ground, the most effective primary treatment used in village drains is roughing filter. It is easy to construct and is low maintenance but of high treatment value in village drains when the focus is just on suspended solids and turbidity. The quality of contaminated surface water is found to be improved significantly when filtered through gravel and sand filters. That is how idea of roughing filter came

TABLE 6.2
Design Velocities of Physical Filters

S. No	Type of Filter	Size Of Media Material	Filtration Rate (m/h)
1	Rock filter	≥ 50 mm	1–5
2	Roughing filter	4–20 mm	0.3–1.5
3	Slow sand filter	0.35–0.15 mm	0.1–0.2
4	Rapid gravity sand filter	1–4 mm	5–15

TABLE 6.3
Media Distribution Pattern in Roughing Filters

S. No	Type of Filter	First Compartment	Second Compartment	Third Compartment
1	Coarse	24–16 mm	18–12 mm	12–8 mm
2	Normal	18–12 mm	12–8 mm	8–4 mm
3	Fine	12–8 mm	8–4 mm	4–2 mm

up. There are two types of roughing filters: vertical and horizontal. Vertical-flow roughing filters operate either as down or up flow filters. Typically, there are three compartments and the size of the media in each compartment decrease in the direction of the flow. The media distribution in practice is given in Table 6.3.

The filter top or at the filter both vertical flow roughing filters incorporate a scum cleaning mechanism and occupies minimal flow when compared to horizontal flow roughing filters. The filter material of vertical-flow roughing filter is completely submerged.

6.6 SECONDARY TREATMENT

The secondary treatment could be biochemical or chemical treatment if primary treatment of physical separation is not sufficient to treat the water to required standards. Depending on the quality of influent and desired quality of the output, an appropriate set of processes need to be adopted. Chemical treatment may be needed if the grey water receives fugitive discharges from tiny-scale industrial activity and/or agricultural fields. Biochemical treatment would be needed if the grey water receives discharges with high organic load like effluents from septic tanks or leach pits or wastewater from tiny eateries in the vicinity.

Technologies adopted for sewage treatment are valid options in these cases. Since the organic load is expected to be much lower than municipal sewage, anaerobic treatment would be adequate in most cases. Exceptional cases with high BOD load had been encountered in which both anaerobic and aerobic treatments were needed to meet the treatment requirements. A detailed description of the technologies for the biochemical treatment is not given here since the same is available in Chapter 5. There are three core groups of technologies available namely, aerobic, anaerobic, and natural. Aerobic processes are generally highly mechanised with high Opex too. The mechanisation gives an advantage in terms of smaller footprints of the treatment plants. Anaerobic systems function in the absence of oxygen. Natural systems are engineered mimics of what are seen in nature like anaerobic ponds, aerobic ponds and constructed wetlands. Due to low pollution loads in village drains, generally such low maintenance technologies are recommended for grey water treatment in villages. Low Maintenance technologies not only keep the Opex low but even avoid the need to employ skilled or semi-skilled manpower and minimise dependence on electric power. As a general principle, only low maintenance technologies need be adopted for grey water management (GWM) in villages and small towns. Low maintenance wastewater treatment systems include mainly anaerobic technologies and natural systems. The anaerobic options include septic tanks, anaerobic baffled reactor (ABR), up-flow anaerobic filters (UAF), and the natural systems include aerobic ponds, anaerobic ponds, and constructed wetlands.

6.7 LOW MAINTENANCE TECHNOLOGIES FOR GREY WATER TREATMENT

The technologies like septic tank, ABR, UAF, and biogas digester are explained in Chapter 5. In practice, when it comes to biochemical treatment of grey water, the most relevant of the technologies for field application would be ponds and constructed wetland. A description of both the technologies is given below.

6.7.1 Ponds

Waste stabilisation ponds are effective low maintenance systems which are effective in treating wastewaters (Droste, 1997). While ponds were among the most popular methods for treatment of municipal wastewaters in earlier part of 20th century, they got replaced by more compact mechanised aerobic systems over a period of time. They use various naturally available microorganisms, algae, etc., for the treatment of the wastewaters. The types of ponds include

- anaerobic ponds
- aerobic ponds
- aerated aerobic ponds
- facultative ponds

In anaerobic ponds, an anaerobic environment is created by having a small surface area and high depth. The depth of such ponds is generally in the range 3.0–5.0 m. The retention time is in the range of 20–50 days. The process here breaks down the organic contents and converts them into methane and other gases. Traces of hydrogen sulphide would be there causing foul smell just like in the case of a Septic Tank or ABR or UAF. Although a very good option for primary treatment for even strong effluents, ideally, anaerobic ponds are used in tandem with facultative ponds and maturation ponds to get best results. Typical of anaerobic reactors, the sludge generation in an anaerobic pond would be quite low.

Aerobic ponds have a large foot print and shall be quite shallow to maintain a reasonable level of dissolved oxygen (DO) for treating the effluents. It is generally used only for weak effluents. Algal growth in the ponds ensures good level of dissolved oxygen by means of photosynthesis. This system is suitable only for weak effluents and generally used as a polishing treatment in the case of sewage. However, for grey water often with low organic load, this is a very good treatment option since it is effectively a solar powered aeration system. Retention time used is generally in the range 15–30 days. Aerated aerobic ponds are aerobic ponds with mechanical aeration systems which increase the efficiency of the process and reduce the footprint appreciably. Facultative Ponds have both aerobic and anaerobic bacteria and depth used is in the range 1.5–2.5 m. The retention time used is ≤30 days. The BOD removal is highest in a facultative pond when compared to other types of ponds at more than 80%.

6.7.2 Constructed Wetlands

Constructed wetland is an option used in treatment of sewage and faecal sludge (CPCB, 2003). Constructed wetlands are referred to as constructed wetlands and root zone treatment systems too. Reed bed, often called planted gravel filter although not exactly a technically correct term, is a type of constructed wetland.

However, the technology has a larger level of application in the case of grey water treatment. Grey water in small towns essentially are the discharges in municipal/village drains which eventually joins a water body or gets used as irrigation water in agricultural fields. Constructed wetlands as well as other low maintenance technologies acquire importance in this context. A wetland is defined as a place where water table is close to or above the top soil level. Natural wetlands had traditionally functioned as sinks for water and wastewater discharged into them. However, the rapid growth in population rendered the concept ineffective and later counterproductive from an ecological perspective. Constructed wetlands also known as root zone treatment system are engineered mimics of natural wetlands for treating different types wastewaters. Since 'engineered', it is a scientific method of wastewater treatment and the first Manual on Constructed Wetlands was published by USEPA in 1988. Although an excellent low maintenance technology, there had been inhibitions in adopting the technology in many developing countries. The Central Pollution Control Board (CPCB), India had published a guideline on Root Zone Treatment System in 2003 but there had been too many inhibitions in adopting this system where availability of land is not a constraint. However, in the case of coastal regulation zones where permanent constructions have some limitations, constructed wetlands has a certain amount of acceptance as a viable treatment option for both sewage and sullage. Sri Lanka is known to have included constructed wetland as an option in treating domestic wastewaters. In spite of being proven technology with a scientific basis for designing and construction as well as availability of manuals and guidelines in various developed and developing countries, it still hasn't gained the popularity it deserves. In any wastewater treatment involving machinery, manufactured hardware and energy inputs are needed which have an ecological impact elsewhere. Constructed Wetland is a green technology since it minimises or even eliminates such requirements.

6.7.2.1 A Brief Description of Constructed Wetlands

Constructed wetlands can be built with a great degree of control with well-defined composites of substrate, type of vegetation and flow pattern. The pollutants are removed through a combination of physical, chemical, and biochemical processes in a reed bed. Settleable and suspended solids need to be removed primarily in a pre-treatment unit, which is to be installed before the constructed wetlands. The suspended solids that remain in the wastewater after pre-treatment are degraded biologically by bacteria attached to the plant and segment surfaces. Nitrogen removal in a wetland is by nitrifying bacterial in aerobic zones to nitrates, which get converted into nitrogen gas by the denitrifying bacteria in anoxic zones. The oxygen requirements for the process are met by diffusion through sediment, water atmosphere interface, photosynthetic oxygen production, and oxygen leakage from macrophytes roots. Constructed wetlands can be used for industrial wastewater treatment in many industries like Food processing, petroleum refineries, textiles, pulp and paper, breweries and distilleries, coal mines, agro-based industries, metal processing, tanneries, etc. (UNIDO, 2001). Constructed wetland has also found application in treating leachates from toxic landfill sites, and garbage dumps.

6.7.2.2 Types of Constructed Wetland Systems

Free Floating Macrophyte-Based System

Free floating macrophyte-based systems uses floating plants with aerial and or floating leaves and well-developed submerged roots. Water hyacinth (*Eichhornia crassipes*) is the most popular variety of plant species used in floating systems. The high productivity of such plants is utilised for treatment of wastewater. In water hyacinth-based treatment systems, while nutrients like nitrogen and phosphorus are removed by incorporation into the biomass of the plants, organic contents of the wastewater get degraded just like in an aerobic reactor. To some extent, nitrification–denitrification too take place in water hyacinth-based systems. The extensive root systems of water hyacinth facilitate high rate of transfer of oxygen from the foliate to the rhizosphere, providing sufficient oxygen for degradation of organic matter and a large surface area for attached microorganism thereby increasing the potential for decomposition of organic matter.

While hyacinth-based systems are quite efficient as a treatment process, the very high retention time in the range of 5–15 days and shallow nature of the system make it comparable to stabilisation ponds in the plant footprint. Further, hyacinth is very vulnerable to salinity which makes the hyacinth-based systems viable only in a limited manner. Hyacinth can be harvested periodically, dried, and shredded for use in biogas plants along with MSW. Since water hyacinth is known to have affinity to absorb heavy metals, the reuse of the plants for composting or use in biogas plants has to be monitored carefully to minimise bioaccumulation of heavy metals. However, in a village scenario with large tracts of free land, the system may be of good use in treating weak grey water.

Rooted Emergent Macrophyte-Based System (Reed Beds)

Rooted emergent aquatic macrophytes (generally referred to as reeds) are dominant life forms in a natural wetland. In general, they produce aerial stems and leaves; they have extensive root and rhizome system. The depth penetration of sediment volume is different for different species. Typical species of emergent macrophytes are common reed (Phargmites), cattails (Typha), and bulrush (Scirpus), All these species can survive in a water-logged condition as well as dry conditions which make them suitable across various climatic conditions and in locations where weather can go to extremes seasonally. These plants work like aerators by transportation of oxygen to the roots and rhizomes. Most such species have an extensive internal lacunal system occupying 50–70% of the plant volume. Oxygen is transported to the roots and rhizomes through gas spaces by diffusion and/or convective flow. Part of the oxygen may leak from the root system into surrounding rhizosphere, creating oxidised conditions in the otherwise anoxic sediments, and stimulating both decomposition of organic matter and growth of nitrifying bacteria.

Emergent Macrophyte-Based with Surface Flow

This is one of the oldest concepts of constructed wetlands. In this case, surface water flow is adopted in narrow and very long ditches planted mostly with the reeds. The presence of submerged portions of the stems as well as the litter is favoured, since they serve as substrate for attached microbial growth. This system has been in use in the Netherlands for over many years.

Emergent Macrophyte-Based with Horizontal Subsurface Flow (HCWL)

This is the most popular version of constructed wetlands. The typical design of an emergent macrophytes-based system with horizontal subsurface flow is a bed planted with *Phragmites australis* or similar reeds and underlaid by an impermeable membrane to prevent seepage. During the passage of the effluent through the rhizosphere of the reeds, organic matter is decomposed biochemically, the anoxic zones within the constructed wetland facilitate nitrification–denitrification process. Nutrients and a part of the heavy metals get absorbed by the plants. Hence adequate care is needed for reusing the trimmings of the reeds for any purpose so as to prevent bioaccumulation of heavy metals. The media used too would adsorb the heavy metals in the wastewater. While treating grey water the possibilities of accumulation of heavy metals are very remote which makes the system safe in the case of grey water treatment and using the trimmings in composting or biogas plants. The functions of the reeds include supply of oxygen to the heterotrophic microorganisms in the rhizosphere of the reeds through the roots and increasing and stabilising the hydraulic conductivity of the soil. The experiences with such systems have indicated effective removal of SS, BOD, COD, nitrogen, and phosphorus. This system is most suitable for sewage treatment in warm countries where the solubility of atmospheric oxygen is lower and dependence on the reeds for supply of adequate quantities of oxygen is greater.

Emergent Macrophyte-Based with Vertical Flow (VCWL)

This along with horizontal subsurface flow system are the two most popular versions of CWLs. Vertical flow constructed wetlands are referred to as French constructed wetlands too. The typical design of this system consists of several beds laid out in parallel with percolation flow and intermittent loading. During the loading period, air is forced out of the soil while the drying period draws air from the atmosphere to fill the pore spaces, thereby increasing the oxygenation. This system ideal in cold countries where the oxygen solubility in water is more. This system is known to be quite effective in removal of BOD, COD, SS, phosphorus, and nitrogen more than the horizontal system and is likely to provide better quality of treated effluent in cold climates. However, in hotter climates, higher levels of bacterial and fungal activity are ensured where a horizontal system will be preferred.

A variant of vertical system is effectively used in treatment of sludge too which works well in any weather as a user-friendly alternative to conventional sludge drying bed. The system operates like a sludge drying bed but has distinct advantages over a sludge drying bed on two counts. The first is that the dried sludge on the top does not inhibit further loading since there would be several hydraulic pathways created by the vegetation. Thus, the frequency of removal of dried sludge is not determined by the quantum of dried sludge on the surface like in a sludge drying bed. The second is that the sludge gets a certain amount of treatment over and above dewatering. A VCWL can be an excellent option for treating faecal sludge due to these factors.

Submerged Macrophyte-Based System

Submerged macrophyte-based systems have their photosynthetic tissue entirely submerged. Submerged aquatic plants are capable of assimilating nutrients from the polluted waters. However, these plants can survive only in oxygenated waters and therefore cannot be used in treatment of effluents with high organic load. These systems can be effective only for polishing treatment and hence have limited practical field application.

Multi-stage System

These consist of a combination of the above-mentioned concepts and other kinds of low maintenance systems. In one of the field cases, a five-star hotel in India, the constructed wetland technology is successfully implemented in the STP as the secondary treatment after anaerobic reactors for sewage and primary settling for sullage.

Constructed wetlands for waste treatment have several advantages over conventional secondary and advanced wastewater treatment systems, which include:

- Low to moderate cost of construction (depending on site conditions).
- Negligible maintenance cost.
- Negligible energy requirement.
- No need of skilled personnel for operation and maintenance of the system.
- Flexibility in handling variations of loading rates.
- Aesthetic appeal.
- Possibility of the system to be operated as a means of income.

The main disadvantages of constructed wetland technology are the need to commit a large tract of land in comparison to mechanised systems. There had been concerns of clogging of the system if the feed have a high content of suspended solids, extreme pH values on either side, and the impact of high TDS/salinity. However, it was found that the impact of these in a CWL was the same or even less in the case of a CWL in comparison to mechanised systems in field cases involving treatment of effluents from tannery industry. Trials with extreme pH taken up as part of a UNIDO project in India in 1990s yielded results alleviating such concerns. Selection of the vegetation acquires critical importance while dealing with industrial effluents since only some of the reeds can survive inundation for long periods as well as dry conditions and extreme fluctuations in pH. Since different species of vegetation suitable for use in constructed wetlands vary in root length, the vegetation needs to be decided upon before designing the CWL. As for oil, grease, grit, and TDS, appropriate pre-treatment needs to be factored in while designing a CWL.

Though the major disadvantage of a constructed wetland is that it requires large area of land compared to conventional system, it is possible to integrate constructed wetlands with the landscape. Thus, constructed wetland is an attractive option of waste treatment for small communities, especially in populated area. The technology has been successfully used across industries over and above treatment of sewage and sullage. It had been proven successful even in cases of some difficult effluents.

6.7.2.3 Impact of Constructed Wetlands

Constructed Wetlands also signify a change in economic thinking. When a nutrient-rich wastewater is discharged its productivity is increased. Where there is a ready technology for harvesting, processing, and marketing the output, this solar power subsidised system can become a better economic competitor than a system not subsidised by solar energy. For example, water hyacinth, one of the fastest growing water plants, can yield 175 tons of high-grade manure and 75,000 m^3 of biogas every year by anaerobic composting from one hectare of sewage pool. Large trees can also be grown in and around constructed wetlands, the wood of which will have a ready market. Use of such vegetation aids treatment of polluted water as well as utilisation of nutrients for increased productivity and value addition. Further, if the net present value of future cash flows is considered along with the capital investment, a CWL has a distinct advantage over other systems. It is also worth noting that unlike in use of other systems, land utilised for setting up a CWL is not lost forever.

Nutrient film technology (NFT) is an offshoot of CWL technology. NFT uses a thin layer of wastewater which flow over the roots of plants. The plants use the water as well as the nutrients in the wastewater for their survival. This method is suitable only for weak wastewater with a very low organic load. NFT is less efficient than a CWL on multiple counts in comparison but has more aesthetic appeal since any plant species can be used in an NFT system. Hence, in cases where large gardens form part of the landscaping, an NFT unit can be a treatment system which would merge with the landscape seamlessly. The limitation of NFT in comparison with a CWL is that NFT ensures good nutrient removal but does not achieve complex treatments which a CWL can. Hence, NFT is suggested only for treating weak domestic sullage or lean sewage in large resorts for reuse in horticulture. A comparison of the key features of low-maintenance technologies for grey water treatment is given in Tables 6.4 and 6.5.

TABLE 6.4
Comparison of Low Maintenance Technologies

Low Maintenance Technologies				
Type	**Type of Treatment**	**Type of Wastewater**	**Advantages**	**Disadvantages**
Septic tank	Sedimentation, floatation Liquefaction of organic solids Conversion of organics into gases Stabilisation of sludge	Wastewater with settleable solids, especially domestic with some organic contents	Simple to construct, durable No requirement of land since located underground	Low treatment efficiency Effluent not colourless and odourless Periodic emptying needed Sludge removed periodically needs further treatment before reuse/disposal

(Continued)

TABLE 6.4
(Continued)

Anaerobic baffled reactor	Anaerobic degradation of suspended and dissolved solids	Pre-settled Domestic and industrial wastewater with narrow COD:BOD ratio	Simple and durable, high treatment efficiency, little space required due to construction underground. Low possibility of blockage	Longer start up time when compared to anaerobic filters
Anaerobic filter	Sedimentation Adsorption Liquefaction of organic solids Conversion of organics into gases Stabilisation of sludge	Pre-settled Domestic and industrial wastewater with narrow COD:BOD ratio	Simple to construct, durable No requirement of land since located underground High efficiency Ideal as a secondary treatment after an ABR or Septic Tank.	Higher Capex due to cost of media Effluent not colourless and odourless Periodic emptying needed Sludge removed periodically needs further treatment before reuse/disposal
Constructed wetlands	Aerobic, anoxic and anaerobic decomposition of pollutants Sedimentation Adsorption Good rates of pathogen removal Low Sludge formation	Suitable for a wide range of effluents-domestic, industrial, etc. Excellent option for difficult effluents	Very high treatment efficiency Aesthetic appeal, extremely low Opex	High requirement of land Cost of Media the highest component of Capex High level of care required while designing and constructing
Anaerobic pond	Sedimentation, anaerobic degradation Sludge stabilisation	Can treat even strong effluents	Simple in construction, Flexible with respect to degree of treatment Little maintenance	High requirement of land Potential for bad odours Possibility of mosquito menace
Aerobic pond	Aerobic degradation, Pathogen removal	Best for weak wastewaters	Simple in construction, reliable in performance, high pathogen removal rates Possibility of fish farming	High requirement of land Possibilities of algae formation Possibility of mosquito menace

TABLE 6.5
Comparison of Technologies for Secondary Treatment of Grey Water for a Basis of 30 m^3 per Day Flowrate

Technology	Minimum Footprint m^2	Capex	Opex	Manpower	Degree of Treatment
Multi stage anaerobic process	25	Mainly Civil works (No land to be committed as the structure can be underground)	No routine Opex Periodical de-sludging needed. Facilities for treatment and disposal of Sludge elsewhere is necessary.	Nil	Up to 90% reduction of organic load
Aerobic pond	200	Mainly Civil Works Large area of Land needed	Nil	Nil	Up to 90% for weak effluents
Anaerobic ponds	60	Mainly Civil Works Large area of Land needed but only about one-third of Aerobic Ponds	Nil	Nil	Up to 90% reduction of organic load
ASP and its variants including extended aeration and SBR	25	Civil Works Equipment for Aeration Pumps	High energy Skilled manpower Routine maintenance	Skilled manpower	Up to 98%
SBR	20	Civil works vary according to site conditions Same technology as ASP, but better efficiency due to superior flow management	High energy	Skilled man power	Up to 98%
SAFF/SPBR	15	Civil works Equipment for aeration Pumps Media Smaller footprint than ASP	High energy Skilled manpower Routine maintenance Occasional cleaning of media needed	Skilled manpower	Up to 98%

(Continued)

TABLE 6.5
(Continued)

MBBR	12	Civil works Equipment for aeration Pumps Media Smaller footprint than ASP and fixed media systems	High energy Skilled manpower Routine maintenance	Skilled manpower	Up to 98%
MBR	18	Civil Works Equipment for aeration Pumps Media Smaller footprint than ASP and fixed media systems Post-treatment sedimentation and filtration may not be necessary	High energy Skilled manpower Routine maintenance Periodical replacement of membrane	Skilled manpower	Up to 98%
Electrocoagulation	20	Very low civil works High value equipment Very low footprint Post-treatment sedimentation, disinfection, and filtration may not be necessary	High energy Skilled manpower Routine maintenance Periodical replacement of sacrificial anodes	Skilled manpower	Up to 99%
HCWL	215	Mainly civil works Large area of land needed Retention time would be about twice that of ASP but due to the low depth, a very large area of land is needed	Practically nil Periodic trimming of reeds and occasional replanting may be needed.	Nil	Up to 96% for weak effluents

6.8 TERTIARY TREATMENT

Tertiary treatment after the biochemical treatment may be needed. Sedimentation may be needed if the technology adopted for biochemical treatment is not CWL. Post-sedimentation, for safe reuse, disinfection too may be necessary in some specific contexts. The processes in tertiary treatment could be:

- sedimentation,
- chlorination with or without dichlorination,
- filtration, and
- membrane filtration in extreme cases.

These processes are covered in Chapter 5.

6.9 TYPICAL TREATMENT SYSTEM FOR CANALS WITH LOW ORGANIC LOAD IN SMALL TOWNS

Treatment system for 'nalas' which feed into a water body like a pond or river will include

- Standard primary treatment units like screening, removal of grit, and grease.
- Rock filter/roughing filter (depending on influent water quality).
- Constructed wetlands (optional).
- Slow sand filtration (if necessary).

6.10 CASE STUDIES AS EXAMPLES WITH DIFFERENT TECHNOLOGIES FOR GREY WATER TREATMENT

In Dungarpur municipality, two treatment systems were set up at a distance of 2 km from each other. In one case, the technology adopted was an HCWL while in the other which had a higher organic load possibly due to the presence of fugitive industrial effluents, the technology adopted was a combination of MBBR and HCWL. There are many lakes within the town, as is the standard practice in Rajasthan. The lakes essentially receive stormwater runoff during rains but do receive water from municipal drains on a regular basis. Of the three major lakes within the town, the municipality desired to improve the quality of water in two lakes. The lakes selected were Gaib Sagar (also written as Gep Sagar) and Suneriya. As a trial a small pilot cum demonstration unit (PDU) was set up in each of the lakes. The two lakes were hardly 2 km apart. However, the effluents reaching Gaib Sagar and Suneriya were having different characteristics. Hence different approaches were needed to achieve the desired results.

6.10.1 Gaib Sagar Lake

The Gaib Sagar Lake in Dungarpur is one of the most beautiful tourist attractions in Rajasthan. The lake has several important temples including the main one dedicated to Lord Srinathji and a few palaces of historical importance. It is a bird watcher's paradise where several unique species of birds are observed. The main lifeline of the town, this lake has deteriorated over a period of time. The lake has lush growth of water hyacinth and algae which get removed periodically using mechanical means. The main reason for this situation is the pollution of the lake by many types of effluents getting discharged into the lake from the town. Most of the drains which discharge wastewater into the lake are connected to a large RCC pipe in the southern side of the lake. There were two drains which were joining a third and the combined stream was the most polluted of the several drains draining into Lake Gaib Sagar. The combined stream was the one having the highest flow rate during summer too. The area between the confluence of the drains and the lake belonged to the municipality and hence it was decided to set up the treatment plant here. Since there was wide variation in the flow pattern throughout the day, a flow management system with an equalisation tank and a facility for pumping at constant rate to the subsequent reactors was necessary. Effectively, this grey water treatment plant resembled a typical sewage treatment plant. A polishing treatment using a constructed wetland, HCWL, too was added towards the end of the plant before the treated water was drained to the lake. Based on the several analysis reports on the quality of the grey water in the drain, a design was prepared which consisted of:

1. Screen chamber.
2. Grit chamber cum grease trap.
3. An equalisation tank.
4. An aerobic reactor using MBBR technology.
5. A secondary settling tank.
6. A sludge holding tank.
7. Disinfection.
8. Filtration through a pressure sand filter and an activated carbon filter.
9. A horizontal flow constructed wetland as polishing treatment.

In this plant, post commissioning, sporadic hydraulic overloading was observed. Although designed to treat 20,000 litres per day (LPD) of effluent but after

FIGURE 6.1 Treatment system for Gaib Sagar water treatment. Photo taken by the author

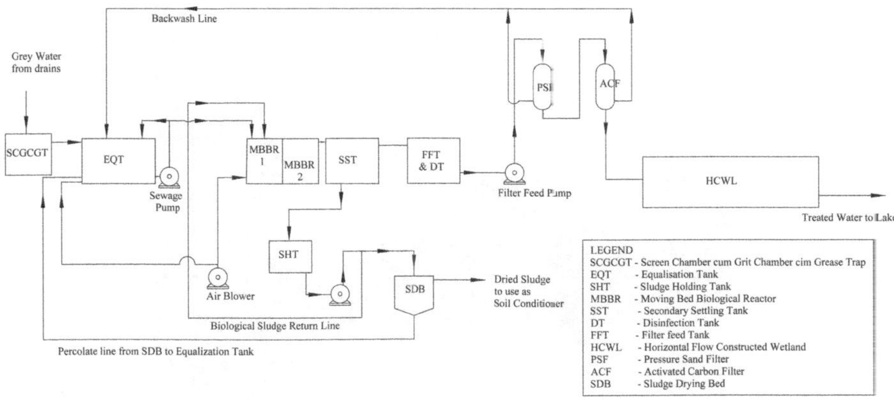

FIGURE 6.2 Process flow diagram for Gaib Sagar Lake

commissioning, started receiving up to 40,000 LPD of effluents. In spite of such a high rate of overloading, the system did not crash and functions smoothly. The CWL originally introduced into the system as a polishing treatment contributed significantly to improve the quality of effluent whenever there was hydraulic overloading (Figures 6.1 and 6.2).

6.10.2 Suneriya Lake

Suneriya was receiving effluents from a set of municipal drains of which a major one was selected for treatment. The municipal authorities wished to have a low maintenance wastewater treatment system (LOMWATS) for treatment of the grey water which was found to be a lean effluent. Accordingly, a suitable system was selected. After monitoring the characteristics of the water in the drain for a certain period of time, it was decided to have a treatment system consisting of the following:

1. Screen chamber
2. Grit chamber cum grease trap
3. An equalisation tank
4. A roughing filter
5. A horizontal flow constructed wetland

The municipality wished to construct the roughing filter and the HCWL inside the lake so that the same can be shown off as working models which could be seen from any point on the road beside the lake. The Capex would automatically be high when the construction is inside the lake but the visual appeal too mattered from the perspective of the municipal authorities.

Although designed for treating 10,000 LPD of grey water, after commissioning, the system got overloaded consistently by 60–100% since the feed got increased to 20,000 LPD. Notwithstanding the higher hydraulic loading, the system was working satisfactorily for several years (Figures 6.3 and 6.4).

FIGURE 6.3 Treatment system for Suneriya water treatment. Photo taken by the author

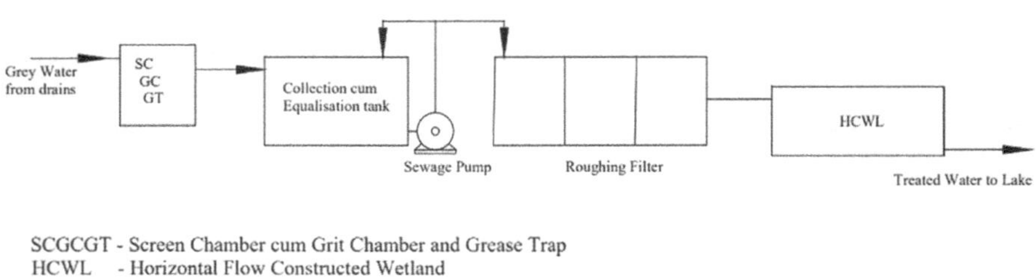

FIGURE 6.4 Process flow diagram for Suneriya Lake

6.10.3 Small-Scale Grey Water Treatment Plant Nandurabar, Maharashtra

The location is a remote village where grey water from a cluster of residences was flowing freely into an agricultural land. There were complaints regarding this and with the support of FINISH Society, a grey water treatment plant was set up. The pollution load was quite low with BOD less than 50 mg/L and suspended solids less than 30 mg/L. The pH was close to neutral and there was no foul smell either. Presence of oil and

FIGURE 6.5 HCWL treatment system for grey water. Photo taken by the author

grease was negligible. There was no need for any highly sophisticated mechanised system due to the low level of pollution load. A set of simple low-maintenance system was designed and installed. The treatment system consisted of

- A screen chamber
- A sedimentation tank
- A roughing filter
- A HCWL

The maximum flow was 9,000 LPD and the flow took place in the morning and in the evening. Due to the low pollution load and practical issues in having a pump, the concept of equalisation tank was dispensed with and the peak load of 2,000 LPD was taken as the flow rate for designing the system. With practically no maintenance except for periodic cleaning of the screens, there was no issue of ownership. The householders linked to the plant took it upon themselves to clean the screen and dispose the screenings to the village-level solid waste management system. The treatment plant yielded very good results with a BOD of less than 5 mg/L and SS of less than 1 mg/L. The treated water was drained to the agricultural field without issues. The farmers and the local people were very contented with the results (Figure 6.5).

6.10.4 Ecological Rejuvenation Project of a Highly Polluted Lake

The strategy adopted for renovation of a highly polluted lake was:

- Study the present condition of the lake and make decision on whether the lake gets a treatment or is left to restore on its own.
- Identify and monitor both the flow pattern and characteristics of the feeding drains.
- Decide on the feasibility of combining drains to minimise the feed points to the treatment facility planned.
- Decide on the kind of pre-treatment each drain requires.

The lake is fed by as many as 22 drains. Each drain had a different feature. There was even some raw sewage carried by a few drains. The lake was highly polluted.

The result was that the entire residential area around the lake was denied groundwater of acceptable quality. There were borewells in the residential apartments and stand-alone residences but practically every well went into disuse. A study of the lake revealed that the lake can rejuvenate itself if fresh organic load is not discharged into the lake. The projected period for self-rejuvenation was about two years. By treating the influents from multiple drains, the flow of organic load and other pollutants to the lake could be cut off. While power was available in the vicinity and the state power board was willing to provide uninterrupted quality power, there was the issue of Opex and ownership of the project. The Capex was anyway grant based but the sustainability of the project was dependent on the system in place for meeting the Opex on a regular basis. The local self-government did have budget constraints. There were some limitations regarding the land too. Striking a balance, a system which is a combination of aerated lagoon in conjunction with HCWL was designed. Different drains could be combined to arrive at as little as four primary treatment units consisting of screen chambers, grit chambers, and grease traps. Due to erratic flow rates, as a means of flow management, an equalisation tank with pumping facility was needed. The outflow from all these were fed to the equalisation tank from where the effluent was pumped to the aerated lagoon provided with surface aerators. The aerated lagoons gave a partially treated effluent which got treated further in a set of horizontal flow constructed wetlands before discharge into the lake. The location was such that during rainy seasons, the system could get inundated since the location in general had such a history. Hence, sufficient protection for the treatment plant from storm water runoffs was provided with adequate arrangements to divert the excess storm water directly into the lake.

The net result of the project is a huge social cost–benefit. The residents were receiving water of good quality from a reservoir several kilometres away by means of tanker trucks. The total wastewater to be treated was in the range 5–7 MLD and the water consumption need of the area around the lake was roughly in the same range. The cost of water transported by tanker trucks for two years was found to be equal to the Capex required for the treatment plant.

Fresh water is becoming dearer over the years, but at the same time, the existing water sources are not managed properly to cater to the fresh water requirements. The approach is to reduce the pollution loads of the inlets and improve the quality of the water. Adoption of low maintenance technologies will ensure sustainability as use of skilled labour, need for power, and maintenance, all of which requires operation and maintenance expenses are liable for system failure. So even if capital expenditure is met and the system is designed and operated, in the long term it may not be sustainable. The quality of groundwater will improve in the long run and the groundwater from wells can be used and can do away with dependence on water transported by trucks from far-off sources. By adopting aesthetic designs by proper landscaping the banks, making recreational parks, walkways, jogging tracks, cycling tracks, boating, fish farming, etc., the nearby residential area will be benefited. Additional benefits are that the removal of water hyacinth will reduce the mosquito menace in the nearby locality and will be a great help to the local residents. The rotten smell from the filth-filled lake can also be done away with.

7 Project Management and Financial Implications

7.1 PROJECT PLANNING, FINANCING, AND ECONOMIC IMPLICATIONS FOR WASTE MANAGEMENT IN SMALL TOWNS

There are four types of waste management systems covered in this book. They are

- Solid waste management
- Faecal sludge management
- Sewage treatment
- Grey water treatment

Broadly, the project planning and financing, contracting, contract management, and project management tools remain the same for all these cases. However, case-specific unique issues are to be addressed separately. The economic viability of each of the four will be based on different concepts. While solid waste management and to certain extent faecal sludge management (FSM) too have an assured regular stream of revenue income, sewage treatment, and grey water treatment need not necessarily have such revenue. The economic viability from a purely commercial angle is doubtful in all cases, but in varying degrees. However, when we factor in the benefits received by society in economic terms, it would justify the investment and Opex in each of the cases. Thus, the responsibility of setting up such waste management systems mainly rest on the society and by extension, the government which represents the people. In cases where there are appreciable economic returns, public–private partnership (PPP) models can be envisaged. A certain amount of support from governmental or corporate social responsibility (CSR)-based funding may be needed as a viability gap funding (VGF) in most of the cases.

7.1.1 Project Planning for Solid Waste Management

Unlike other types of waste found in a town, solid waste is a heterogeneous waste with several components. The solid waste management (SWM) involves more manpower for collection, segregation, transportation, processing, and marketing of products. As explained in Chapter 3 of this book, the solid wastes in a town will have components which can be sold directly, which can be sold after some value addition and some which need to be disposed safely without affecting the environment. In view of the possibility of operating a SWM profitably, public–private partnership (PPP) models are possible. Currently, the field cases cited in this book indicate an initial investment by the government in terms of land and infrastructure. The economies of scale come

DOI: 10.1201/9781003650737-7

in to play when it comes to processing and value addition of solid wastes. The project planning starts with:

- Generation of data on population, both resident, and floating.
- Quantification and characterisation of wastes.
- Having a protocol for the citizens on the means of disposing the wastes.
- Developing a strategy for segregation.
- Identification of markets for saleable components of MSW.
- Exploring possibilities of development of entrepreneurs.
- Deciding on processing options.
- Identification of an ideal place to locate the processing plant.
- Identifying potential partners if the SWM is proposed as a PPP model.
- Assessment of viability gaps if any when it comes to Opex.
- Local self-government level enforcement of laws regarding SWM.

Once these steps are covered, the feasibility needs to be studied and a detailed project report (DPR) need to be prepared. The precise figures regarding Capex, Opex, manpower requirements and need for replacement of machinery and the time frame for the same need to be included in the DPR. The DPR will cover the environmental impacts and management plans. From the DPR, the commercial viability of the project can be assessed and financing of the project can be organised accordingly. As had been explained in Chapter 3, the revenue streams could sustain the SWM operations. The field case of Balangoda shows how entrepreneurship can be developed and jobs can be created. The case of Ponnampatty shows how the SWM centre can be made aesthetically appealing and used as a place of leisure and entertainment by the public. The SWM can have a wide variation from place to place, as already explained. Hence, the treatment plants to be constructed too can vary. The basic common threads are transportation of solid wastes to the processing centres and a composting unit in the processing centre. For transportation, the current trend is to use electric vehicles. Irrespective of the type of vehicles used adequate number of compartments are to be provided in the vehicle for keeping the segregated materials separate from one another. Since only open-source technologies are recommended for small towns, the composting method adopted would be aerobic windrows method which entails construction of the shed and facilities for draining of leachates and a treatment plant for the leachates. The cost of construction of the composting shed is dependent on the soil conditions. The cost of land, vehicles for transportation, and the plant with machinery like pulveriser, shredder, and stitching machine are one time investment which need to be supported by the investor which usually is the government or the local self-government.

In case the operations are run in a PPP model, the private entrepreneur may bring some funds at this stage of the project. The standard commercial analysis of indices like internal rate of return, return on investment, return on equity, payback period, and debt service coverage ratio need to be calculated and reviewed after eliminating any VGF for assessment of the feasibility of the project in a PPP model. If involving NGOs or Workers' Cooperatives, the feasibility analysis would throw up similar results but with a different impact. Since the subject is waste management per se, the Economic Rate of Return (ERR) assumes significance in this case. The indirect social and environmental benefits from a SWM scheme will contribute significantly to the Economic Rate of Return.

Construction is a multi-dimensional affair in the case of SWM. The construction of Receiving cum Sorting Centre both in the town and the Processing Centre/Resource Recovery Park, Construction of sheds for composting and sales outlets are the essential buildings needed. Beyond this, there would be purchases of vehicles and machinery. Tendering is the safest and the most practical route unless the purchases or constructions are funded by a CSR fund or by an NGO. Tendering means, there should be detailed design and technical specifications and bill of quantities (BoQ) for each item. Since different types of works form part of the project, there will be several contractors and suppliers involved. To smoothen the project execution there should be a coordinator either from within the government or an external specialist agency assigned for the purpose. Beyond the construction, purchases, commissioning, etc., the government has an additional responsibility of creating awareness in the general public as well as the workers dealing with the waste. As can be seen in Chapters 3 and 4, a commercially oriented faecal sludge management (FSM) system ideally gets integrated with SWM. Hence, procuring equipment and constructions with respect to FSM too will become part of the SWM Resource Recovery Parks. These activities too need to be integrated with the SWM while planning the whole scheme. The integrated scheme with multiple products like fortified compost/co-compost, vermi compost, vermi wash, plastic ingots, HDPE granules, and plastic impregnated concrete blocks is likely to get investment partners if conceived as a PPP project. Strict monitoring of movement of faecal sludge can ensure that the FS reaches only the processing centre and that in turn creates a revenue stream of tipping fees too adding to the commercial viability of the project. FSM, other than being an essential environmental service to be provided by the government also has the potential of being viable and generating revenue from one or all of the following activities:

- Tipping fees for the private operators for sludge disposal by the local bodies
- Sale of co-compost/fortified manure

7.1.2 Sample Economic Analysis for a Co-Composting Plant

A sample economic analysis of a typical co-composting plant in India with degradable MSW and FS as raw materials and two processes of treatment of FS and co-composting of partially treated FS with MSW is given below. The Capex for the project was INR 1,000,000 which was met by the local self-government with the support of an NGO.

The investment in setting up the FSTP and its operations and maintenance will vary based on the choice of technology and capacity of the plant. The return on the investment made yields indirect and intangible benefits to the society which includes a cleaner environment and better living conditions in the town over and above the visible returns in cash terms as indicated in Table 7.1. In order to achieve long-term sustainability for an FSM plant, it is critical that there are:

- Strict regulations in place and enforced for the private players for tipping fees and penalties.
- Local government dedicates land for FS disposal, treatment, and processing.
- Initial investment is secured through a grant or invested by the government.
- Adequate resources are dedicated to O&M of the FSTP.
- Technical support from a field expert on suitable technology or design.
- Regular monitoring from an independent partner.
- Social acceptability and behavioural change from citizens.

TABLE 7.1
Economic Analysis of a Typical Co-Composting Plant in India

S. No	Means of Income	INR/Month	INR/Year
1	Sale of co-compost (3.75 T/month at INR 5/kg)	18,750	225,000
2	Tipping fee (3 trucks per day for 25 days a month)	22,500	270,000
3	Total revenue	41,250	495,000
4	Opex	35,000	420,000
5	Net Profit	6250	75,000

There would be exclusive FSTPs which hardly can generate revenues other than tipping fees since the end product can at best be a soil conditioner with little commercial value. In case of technologies like pyrolysis, the end products like combustible gases do have a monetary value, but economies of scale become crucial in such cases. If it is possible to combine the FS from several small towns into a centralised pyrolysis plant, that would still be a difficult proposition since the transportation costs of raw FS would be a burden on the operators of hauler trucks which in turn would be passed on to the customers.

7.1.3 The Responsibilities of Each Stakeholder Involved in a Typical PPP Model

The responsibilities of each stakeholder involved in a typical PPP model for an integrated SWM-FSTP system would be as follows.

7.1.3.1 Government

- Regulations
- Allocation of land
- Support for Capex
- Support for replacement of machinery needed.

7.1.3.2 Government/NGOs

- Technical support
- Capacity building
- Grant support
- Social awareness and behavioural change
- Post project monitoring.

7.1.3.3 The PPP Partner

- Collection of tipping fees
- Sale of products
- Opex

7.2 PROJECT PLANNING AND IMPLEMENTATION FOR SEWAGE TREATMENT

There are two types of sewage treatment plants under consideration here. One is a centralised sewage treatment plant for a part or even the complete town. Other is an STP within a commercial building like a hotel, resort, or shopping mall. The captive STP will yield treated effluent which may be reused within the building or disposed to soil by means of soak pits or dispersion trenches or seepage beds. The financial considerations in the latter can be considered an internal issue of the commercial institution and hence do not need a detailed consideration here.

A centralised municipal sewage treatment plant would be required as an off-site sanitation system to eliminate the hassles of maintenance issues in on-site sanitation systems like septic tanks and leach pits. The treated sewage can have various applications like make up water in cooling towers, raw water for process industries, irrigation water for agriculture, or simple recharging of groundwater. The degree of treatment required would be based on the end use/means of disposal. Setting up a sewage treatment plant and the associated sewer system would involve planning, contracting, project implementation and post project operation and maintenance.

7.2.1 Capacity and Site Selection

For setting up a sewage treatment plant, we need to know, to start with, the capacity based on the current population and projected population of the local body under consideration. This will enable projecting the capacity of the treatment plant required. The site selection is the next step. This will depend on the land available with the local body. If such a parcel of land is not available, the possibility of acquiring/purchasing a piece of land at minimum cost with a technically and legally viable means of disposing or reusing the treated sewage needs to be identified. The viability of laying sewers in the town and transporting the sewage to the project site too is of immense importance. The site selection process and assessment of soil quality have potential direct and indirect environmental impacts too. The characteristics of the soil will affect the cost of construction since it directly impacts the structural design and the materials of construction. Environmental impact can be direct and indirect. Environmental impact assessment needs to be taken up and a properly designed environmental management plan needs to be in place during the planning stage itself.

7.2.2 Sewer Network

In order to transport the raw sewage to the sewage treatment plant, a sewer network needs to be planned. The area under the local body has to be mapped to assess the possibilities of minimising pumping stations by selecting the routes according to the slope and direction of flow of raw sewage. Minimisation of pumping stations means minimisation of equipment which in turn implies serious reduction in O&M costs. In many case, the beneficiaries of the sewer network and the STP are charged a connection tax. Such citizens need to be taken into confidence and there should be consensus regarding the financial implications to the citizens. Necessary traffic management systems too should be planned well in advance with the involvement of the stakeholders before laying of sewers is planned and implemented.

7.2.3 Selection of Technology

Selection of technology needs to be based on multiple factors, like

- Availability of land
- Means of financing the Opex
- Skilled labour
- Power
- End use patterns or means of disposal
- Climate impacts

Generally, the footprint of the plant is almost inversely proportional to the Opex. Lower the operation and maintenance costs, higher the requirement of land is the general principle. Financing of the Opex acquires critical importance. If the treated sewage is sold for a price, at least a part of the Opex can be recovered from the sale proceeds. Likewise, the connection tax paid by the beneficiaries ensure a regular stream of revenue income. Availability of power and a reliable source of power will influence the selection of technology. In a mechanised aerobic system, the operation needs to be continuous and the power supply should be uninterrupted. The pumping stations located in various parts of the town too should have access to power. Even when uninterrupted power is available, the general concept is to have a backup generator to meet any exigency. Lack of access to power or a power source will narrow down the choices of technologies.

End uses can vary from reuse for irrigation or industrial purposes and disposal can be to soil or a water body. The standards for disposal and reuse vary from case to case and the degree of treatment too will vary. This factor can influence the selection of technology. For example, although only limited treatment is required for disposing the treated sewage to an ornamental pond with or without an aesthetic fountain, the aesthetics will influence the selection of the treatment process. In such a case, clarity of the treated sewage is of paramount importance; so is the odour factor which necessitates a very good degree of removal of odour and turbidity, while the organic load is not exactly relevant. If the end use is for a captive heat exchanger, additional treatment for removal of hardness would be needed. The feasibility of disposal into soil depends on the characteristics of the soil too which in turn will influence the decision-making regarding selection of technology. For example, a natural system like constructed wetland is a green technology with minimal impact on climate, not requiring skilled operators and practically eliminates the need for sludge management, requires commitment of a large area of land. On the other hand, a sophisticated and proven technology like electrocoagulation has a very small footprint but needs high quantity of electrical energy and has a high O&M cost in terms of periodical replacement of the sacrificial electrodes.

7.2.4 Preparation of Detailed Project Report

Once the location, capacity, and technology are finalised, a detailed project report need to be prepared. In some cases, a feasibility report may precede the detailed project report. This may be necessary to have a fund allocation for the Capex and Opex of the project. This would indicate the Capital Expenditure (Capex) with detailed bill of quantities, means of financing, energy requirements, labour requirements, cost of operation and maintenance (Opex) and possibilities of revenue income from sale of treated sewage. It is at this stage that the means of financing the Capex and Opex need to be identified. Unlike in the case of a solid waste-FS processing plant, the possibilities of a PPP model would be possible only if the final product, the treated sewage has a sale value which can cover at least the Opex.

7.2.5 Legal and Environmental Compliances and Aesthetics

For setting up a sewage treatment plant, there will be several legal clearances required based on the local laws. Before going into the contracting stage, these steps are to be taken. Irrespective of the technology adopted, adequate green cover is to be planned for the treatment plant area. The green cover offers multiple benefits. If the treatment plant uses aerobic process, there would be emission of carbon dioxide which can be absorbed by the trees in the area. Adequate number of trees can be planned well in advance based on the projected generation of carbon dioxide. Further, the green cover provides protection for the neighbourhood from any fugitive odours. The treatment often generates some sludge fit for use as soil conditioner. Instead of spending time and money to dispose the sludge, a small flower garden can be maintained within the green cover where the sludge can be used.

7.2.6 Selection of Implementing Agency and Contracting

This is done by different methods. The most popular method is to seek tenders based on the BoQs in the DPR. This will require minimal effort in scrutinising the tenders. The bidders need to meet the prequalification requirements in terms of experience, financial capability, and availability of skilled professionals for the execution of the project. Thus, the scrutiny is limited to the financial aspect alone.

Alternate method is to have design tenders. In this case, the bidder has to prepare the design and quote as per their own design. For scrutiny of such tenders, a technical committee of skilled and experienced professionals is needed. This method often is a minefield since each of the bidders may come up with different technologies and systems, making the comparison extremely difficult. In order to reduce such complications, the technology to be adopted can be mentioned in the tender document.

7.2.7 Construction Stage

The construction stage of the sewage treatment plant includes

- Civil works like tanks, office buildings, laboratories, etc.
- Purchase and installation of mechanical equipment.
- Providing appropriate electrical connections.
- Post project trial runs.

Each component will need strict quality control and assurance. In the contracting stage itself, a detailed manual on quality control and quality assurance need to be prepared and handed over to the contractor. The agency overseeing the construction and Installation should ensure compliance of the stipulations in the manual and adhere to the predetermined standard procedures. Likewise, the safety, health, and environmental aspects need to be addressed in the project implementation stage. A manual for SHE too needs to be prepared for this purpose and the agency overseeing the project implementation should ensure adherence to the guidelines and stipulations. Timely completion of the project is critical from multiple perspectives. A proper schedule needs to be prepared and adhered to. Time overruns and cost overruns need to be avoided. Laying of sewer lines, connections to manholes from sources of wastewater and location of pumping stations for part of the sewage collection and transportation part of the project. Necessary clearances from the appropriate authorities would be needed for this work too since the sewer network would essentially be in a public space. Availability of continuous power supply for the pumping stations needs to be ensured, safety measures ensuring that the pumping stations are accessible only to authorised personnel is quite important.

7.2.8 Trial Runs and Commissioning Stage

Identification of operators for the operation and maintenance of the plant and imparting training to them is needed at this stage. Along with this, a Manual for Operation and Maintenance needs to be prepared. The log sheets for use in the operational stage need to be finalised at this stage. Arrangements for disposal of hazardous wastes from the sewage treatment plant need to be made at this stage. Once the operation commences, such issues should be handled as a matter of routine by adopting standard procedures. Before handing over, it would be ensured that the plant and the transportation system function smoothly. Once the sewer network is ready to transport the municipal wastewater to the sewage treatment plant and the plant is ready to function, the initial days of operation would require close performance monitoring. The quality of the treated effluent should meet the predetermined specifications. At this point, the disposal or reuse of the treated effluents would commence.

7.2.9 Economic Implications

If there is a buyer for the treated wastewater, the sale should be monitored as well as the quality of the treated water sold. Metering of the sale and disposal are to be done separately and accounted separately. The proceeds from the sale should be used to partially finance the Opex. In exceptional cases, the entire treated sewage is sold off and, in such cases, the plants have switched to PPP model investment due to the high level of profitability with an assured stream of continuous income. The intangible benefits to society are a cleaner environment. The beneficiaries who pay a connection tax are saved from hassles like

- Maintaining an on-site sanitation system like Septic Tank
- The difficulties encountered for arranging periodic emptying of the substructure
- Lack of access to the washrooms during the emptying operation

The public is protected from any kind of potential nuisance when an FS-laden truck is driven through the town at odd hours.

7.3 GREY WATER TREATMENT PROJECT IMPLEMENTATION

This is as complex as SWM but with different dimensions. Grey water treatment plants, unless if not planned as part of a restoration project of large water body, are generally much smaller than STPs. Treated grey water normally do not ensure a direct revenue income unless if it is located in water starved area with industrial needs for water. The reference in sewage treatment like reuse for industrial applications is valid for grey water too. The factors which affect the management approach would depend on:

- Characteristics of the raw grey water
  Unlike sewage, characterisation needs to be a priority in this grey water management. There could be multiple streams, each with different features, joining at a common point, a receiving water body. This becomes critical when the objective of grey water treatment is to improve the quality of water in the receiving water body.
- Quantity to be treated
  Quantity of grey water to be treated determines the technology options and land requirements.
- Location of the treatment plant
  Location of the plant has direct impact on the selection of technology since there are social concerns, potential for public nuisance like unacceptable odours and aesthetics.
- End use/means of disposal
  End use generally is either to agricultural fields or to a water body. Grey water generally is not disposed into soil by means of soak pits or dispersion trenches.

The grey water treatment plants invariably are grant-based models supported by governments or NGOs or corporate CSR funds. The investment yield returns only to society and not directly to the investor. Lake rejuvenation programmes can have considerable value addition by developing places of leisure and entertainment around the water body. It would be far more attractive to the general public than the parks located in RR Parks and sometimes generate some income for the society and the local government too. Notwithstanding the lack of economic viability, grey water treatment has a huge positive impact on the environment and the society. The control of pollution is what is observed first. But over a period of time the quality of ground water would improve which reduces mining and transportation of fresh water from elsewhere. The transportation of water from distant locations has a big ecological footprint. The local population which was denied the opportunity to use groundwater due to its poor quality gets better groundwater which has a direct economic impact on the society. Further, if the lake were a fishing area, there would be a livelihood impact too by treating grey water to improve the quality of the water body.

TABLE 7.2
Economic and Socioeconomic Impacts of Waste Management

S. No	Type of Waste	Capex	Opex	Socioeconomic Outcome
1	Solid waste	Grant-based or PPP or a mix	Can be covered from revenue income in both grant-based and PPP models	• Employment generation, Development of Entrepreneurship • Cleaner environment • Multiple products • Resource conservation
2.a	Faecal sludge	Grant-based	Grant-based	• Soil conditioner as product with low sale value • Cleaner environment
2.b	Faecal sludge with co-composting	Grant-based or PPP or a mix	Can be covered from revenue income in both grant-based and PPP models	• Co-compost or fortified manure as final product generates income • Circular economy • Cleaner environment
3	Sewage	Grant-based	Grant-based unless the treated sewage has a sale value	• Treated sewage which may have a sale value • Elimination or reduction of in situ sanitation systems and associated social and economic implications • Cleaner environment
4	Grey water	Grant-based	Grant-based	• Indirect benefits to society by improving quality of groundwater • Good quality water for irrigation • Reduction of pollution in water bodies • Cleaner environment

7.3.1 A Cost–Benefit Analysis of a Typical Case of a Lake Restoration Project with a Grey Water Treatment Plant of Capacity 7 MLD Is Given Below

The Capex committed was INR 28,200,000. The rejuvenation of the lake would result in improving the quality of the groundwater. The residents were expected to use 3.5 MLD in the second year and 4.2 MLD in the third year and 4.9 MLD from the fourth year onwards. Assuming a nominal rate of INR 5 per m^3 of water supplied by tanker trucks from distinct sources, the value of 3.5 MLD of water would be INR 1,178,000 in the second year which would be 6,655,800 in the third year, 7,888,880 in the fourth year, and INR 9,129,968 in the fifth year onwards. Over and above this direct benefit to the society, there would be an equivalent savings of water at the source. The ecological footprint of over 400 truck trips plying everyday too gets eliminated by the lake restoration project of which the core activity is grey water treatment.

The socioeconomic impacts of waste management are summarised in Table 7.2.

7.4 CONTRACTING, CONTRACT MANAGEMENT, AND PROJECT MANAGEMENT

The projects in each case involve construction, supply, installation, testing, and commissioning of equipment. The governments or NGOs involved in waste management projects will not have own capabilities for taking up the complete implementation of the construction works or the supply and installation of equipment. Hence, it is necessary to engage specialist agencies who can take up such works professionally. Selection of contractors starts with prequalification of contractors in both technical and financial capabilities. There basically are two methods of contracting in grant-based models, namely design build operate (DBO) or work allotted on the basis of bill of quantities. In the case of PPP model or a grant-PPP mixed model, there are multiple ways of designing the contract. The standard methods include build own operate (BOO), build own operate transfer (BOOT), and build operate transfer (BOT). In such models, the contractor may bring in the technology too if necessary. In BOO models, the private partner owns and operates the treatment plant without any time limit. It is an open-ended arrangement. In BOO models, the grant may be only the land allotted for building the plant. The BOOT model is similar to BOO, but the ownership of the project is only for a finite period after which the plant is transferred to the government. The government would be free to select an operation and maintenance (O & M) contractor for operating the plant for a fee or engage a contractor who can take the full responsibility while keeping the proceeds from sale of products from the plant.

In BOT models, the private partner operates the plant for a predetermined finite period after which it is transferred to the government. The government may engage another agency to operate the plant or operate it directly.

The project management in the case of sewage treatment plants and grey water treatment plants use the standard tools to manage the time frame for construction, maintain the financial outlay within the budgeted amount along with the procedures for quality control and quality assurance of construction (Figure 7.1 and Table 7.3). Prevention of cost and time overruns without dilution in quality of construction makes a project a real success. Post-project monitoring of the performance of the plants form part of the project itself since achieving the treatment objectives is the core in evolving a project. In grant-based models, exploring new avenues of revenue income too need to be explored in post project situations. The field cases of Ponnampatty and Busia referred to in Chapters 3 and 4 indicate how revenue income can be increased by cultivating high-value products and value-added agricultural products like saplings and seeds within the RR Parks and sell them in the market.

TABLE 7.3
List of Project Activities and Reference to Activity Network Diagram for Lake Rejuvenation

Activity No	Activity	Node of Activity	Duration (days)
1	Obtaining clearances from government	101-301	28
2	Floating tenders	101-201	5
3	Receiving tenders for construction, evaluation of the same, and award of the work	201-301	15
4	Reclamation of parts of the lake wherever it is required for construction of treatment plants by using the dredged materials, levelling, compacting, and testing	301-302	20
5	Preparation of land for construction of the composting units	302-401	10
6	Ordering for materials of construction	401-502	5
7	Delivery of construction materials at site	502-601	20
8	Ordering for mechanical and electrical components of the treatment plants and composting units	301-507	5
9	Delivery of equipment and components at the site	507-601	40
10	Applying for power connections	101-508	2
10A	Obtaining power connections	508-601	30
11	Identification of sources of reeds and preparatory work for collection and transport	301-501	5
11A	Developing a mini nursery for reeds including preparation of the site, collecting the reeds from sources, trimming, and transportation to the site	501-601	20
12	Site preparation for treatment plant of Inlet 4	401-503	4
15	Civil construction for treatment plant of Inlet 4	503-601	40
13	Site preparation for treatment plant of Inlet 14	401-504	8
16	Civil construction for treatment plant of Inlet 14	504-601	60
14	Site preparation for treatment plant of Inlet 17	401-505	15
17	Civil construction for treatment plant of Inlet 17	505-601	75
18	Site preparation composting units	401-506	10
19	Civil construction of composting units	506-601	30
20	Installation of equipment and planting of reeds	601-602	8
21	Trial runs and commissioning	602-603	20

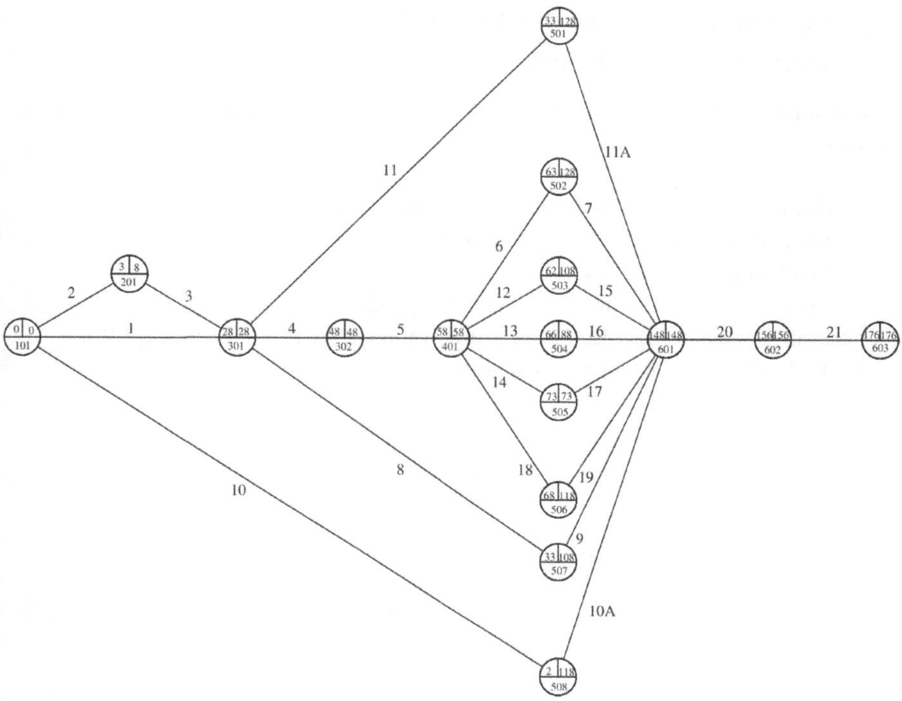

FIGURE 7.1 PERT chart lake rejuvenation project

7.5 CONCLUSIONS FROM THE SUCCESS STORIES

- Municipal solid wastes can be used to manufacture various tradable products and not just energy and compost.
- Integrated approach to waste management can lead to a system which is viable both from technical and commercial angle.
- This indicates that adoption of appropriate methods and systems for waste management cannot only improve the quality of life of the citizens but also act as a means of income for the local body.
- Addressing many livelihood issues successfully is an added bonus from appropriate waste management systems.
- The sustainability of waste management in any context depends on a wide variety of factors. One critical aspect is the economic viability. While Capex can be met with funds from one or other source, meeting Opex in grant mode has inherent risks. It has been seen that recovering the Capex is almost impossible in the case of small towns. However, there are several success stories where the Opex has been covered with the revenue income from waste management.

7.6 INHERENT RISKS IN SUSTAINABILITY OF WASTE MANAGEMENT SYSTEMS

The sustainability of waste management systems depends on multiple factors which include

- Commitment and capability of the administrators
- Interest and support of the political leadership
- Involvement of the society
- Fluctuations in the market for saleable products
- Innovations and constant upgrading of the products and product profile
- Technology obsolescence in specific cases

References

Anand, Y. P., 2016. Cleanliness – Sanitation: Gandhian Movement and Swachh Bharat Abhiyan. Available at http://www.mkgandhi.org/articles/cleanliness-sanitation-gandhian-movement-swachh-bharat-abhiyan.html. Accessed on 10/9/2025.

CPCB, 2003. *Guidelines on the Construction, Operation and Application of Root Zone Treatment Systems for the Treatment of Municipal Industrial Wastewater.* Central Pollution Control Board, Government of India, New Delhi.

CPCB, 2016. *Municipal Solid Waste (Management and Handling) Rules 2016. The Gazette of India: Extraordinary, Part II Section 3(ii).* Ministry of Environment, Forests and Climate Change, Government of India, New Delhi [pdf]. Available at https://cpcb.nic.in. Accessed on 5/9/2025.

CPHEEO, 2016. *Municipal Solid Waste Management Manual*, Central Public Health & Environmental Engineering Organization, Ministry of Urban Development, Government of India, New Delhi. Available at https://mohua.gov.in/upload/uploadfiles/files/Part2.pdf. Accessed on 5/9/2025.

CPHEEO, 2020. Composition and Quantity of Solid Waste. mohua.gov.in. Available at https://www.cpheeo.gov.in/upload/uploadfiles/files/chap3.pdf. Accessed on 15/9/2025.

CS, 2019. Solid Waste in India. Available at https://csestore.cse.org.in/pub/media/catalog/product/file/sample-to_burn_or_not_to_burn.pdf. Accessed on 2/9/2025.

CT, 2022. Climate Transform, What Role Do Trees Play in Today's Global Warming. Blog, Climate Change, Tree Planting. Available at https://climatetransform.com/what-role-do-trees-play-in-todays-global-warming/. Accessed on 6/9/2025.

Davis, M. L.,2014 Mackenzie L.Davis, David A. Cornwell, *Introduction to Environmental Engineering*, Fifth Edition, McGraw Hil Education(India) Private Limited, New Delhi.

Droste, D. L., 1997. *Theory and Practice of Water and Wastewater Treatment*, John Wiley and Sons, New York.

EPA, 2014. Requirements for Landfills. Available at https://www.epa.gov/system/files/documents/2024-01/ar-rule-19-chap-17-part-62-111d-requirements-for-landfills-pages-from-epa-r06-oar-2022-0546-0003.pdf. Accessed on 10/9/2025.

FAO, R. V. Misra, M. N. Roy, H. Hiraoka, 2003. On Farm Composting Methods. Available at https://www.fao.org/4/y5104e/y5104e06.htm. Accessed on 4/9/2025.

FS, A. J. James, G. Anand, Yugandhar Mandavkar, M. Subburaman, Dhruv Athreye, 2017. Learning Guide for FINISH Society Solid and Liquid Waste Management (including Faecal Sludge) in Rural Areas 2017, FINISH Society and FINISH Mondial.

Hammer, D. A., 1989. *Constructed Wetlands for Wastewater Treatment*, First Edition, CRC Press, Boca Raton.

IS, 1979. IS 9235(1979) Physical Analysis and Determination of Moisture in Wastes (Excluding Industrial Solid Wastes) [CHD 32: Environmental Protection and Waste Management]. Available at https://law.resource.org/pub/in/bis/S02/is.9235.1979.pdf. Accessed on 1/9/2025.

Metcalf and Eddy, 2014. *Wastewater Engineering Treatment and Resource Recovery*, Fifth Edition, McGraw Hill, New York.

MoEF, 2016. Solid Waste Management Rules, Ministry of Environment and Forests, Government of India. Available at https://www.iwma.in/HWM%20Rules.pdf. Accessed on 3/9/2025.

Naresh, 2020. General Environmental Acts. Available at https://data.vikaspedia.in/short/lc?k=CMPaR-iLRc4PdcSi14Os6w. Accessed on 12/8/2025.

Prescod, M. B., 1992. *Wastewater Treatment and Use in Agriculture.* FAO Irrigation and Drainage Paper 47, Food and Agriculture Organization, Available at http://www.fao.org/docrep/T0551E/t0551e00.htm#Contents. Accessed on 7/9/2025.

Rajagopalan, R., 2015. *Environmental Studies from Crisis to Cure*, Third Edition, Oxford University Press, New Delhi.

Strauss, M., 1996. 'Health (Pathogen) Considerations Regarding the Use of Human Waste in Aquaculture'. *Environmental Research Forum* Vol. 5, pp. 83–98.

Tchobanoglous, G., 2019. George Tchobanoglous, Hillary Theisen, Samuel A. Vigil. *Integrated Solid Waste Management*, Indian Edition, McGraw Hill Education, Chennai.

UNIDO, 2001. *Report on Reedbeds for the Treatment of Tannery Effluent, 'Regional Programme for Pollution Control in Tanning Industry in South East Asia'*, UNIDO, Vienna.

UNFCCC, 2024. United Nations Climate Change, September 2024 Food Loss and WASTE. Available at https://unfccc.int/news/food-loss-and-waste-account-for-8-10-of-annual-global-greenhouse-gas-emissions-cost-usd-1-trillion#:~:text=significant%20environmental%20burden.-,Food%20loss%20and%20waste%20account%20for%208%. Accessed on 1/9/2025.

WHO, 1989. *Health Guideline for the Use of Waste Water in Agriculture and Aquaculture.* Technical Report No. 778, WHO, Geneva.

WHO, 2006. *Guidelines for the Safe Use of Wastewater and Excreta in Agriculture and Aquaculture, Measures for Public Health Protection*, World Health Organization, Geneva. Available at https://www.who.int/teams/environment-climate-change-and-health/water-sanitation-and-health/sanitation-safety/guidelines-for-safe-use-of-wastewater-greywater-and-excreta. Accessed on 12/9/2025.

Index

I

M

N

O

P